RARE PLANTS

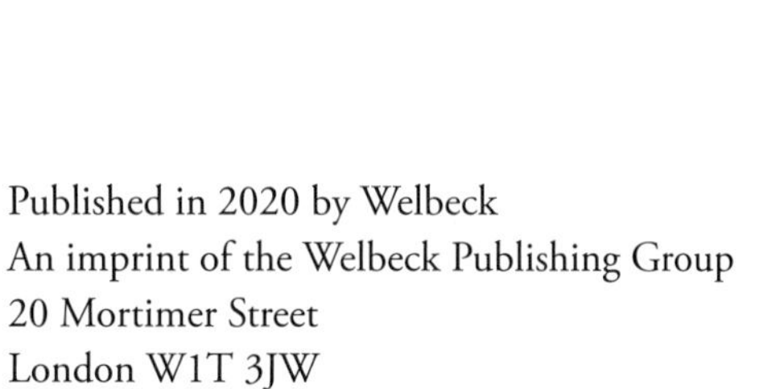

Published in 2020 by Welbeck
An imprint of the Welbeck Publishing Group
20 Mortimer Street
London W1T 3JW

A CIP catalogue for this book is available from the British Library.

ISBN 978-0-23300-623-9

Printed in China.

10 9 8 7 6 5 4 3 2 1

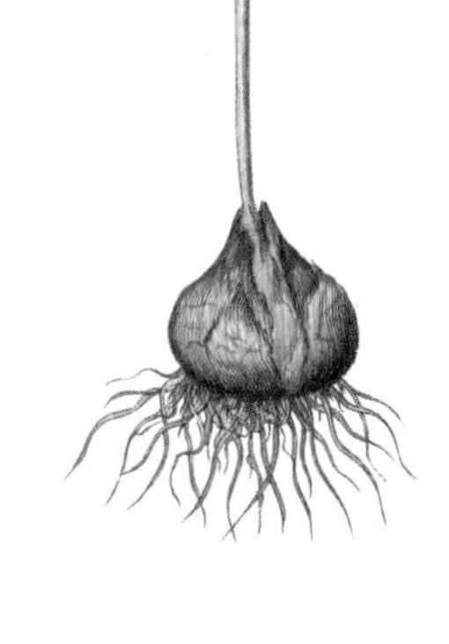

Royal Botanic Gardens Kew

RARE PLANTS

The story of 40 of the world's most unusual and endangered plants

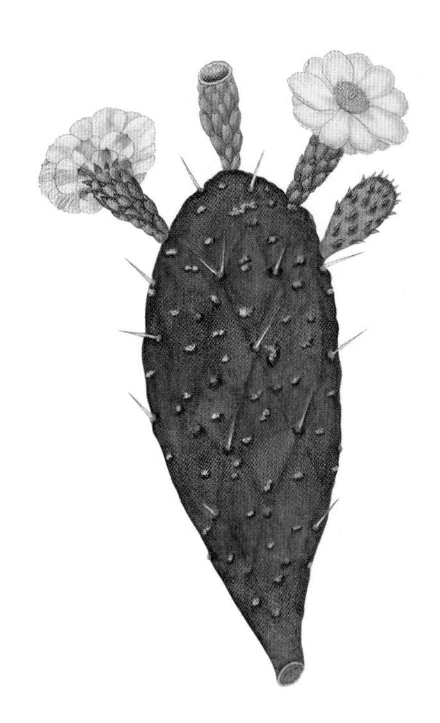

ED IKIN

WELBECK

Nepenthes distillatoria *fœmina*.

ACKNOWLEDGEMENTS

To the amazing Kew publishing team: Lydia, Pei and Gina for their research, expertise and support.

To my expert Kew science and horticulture colleagues for their generosity of knowledge: Dr Paul Wilkin, Dr James Borrell and Dr Aaron Davis on all matters Ethiopian, Dr Olwen M Grace on aloes and new research methods, Dr Michael Way on UK plants that should not be forgotten, Dr Justin Moat for demystifying remote sensing, Dr Kate A Hardwick on restoration, Tim Pearce for shedding light on Africa's *Streptocarpus*, Dr Elinor Breman and Dr Colin Clubbe for expert conservation insight, Iain Parkinson and Jo Wenham for meadow and propagation knowledge, Richard Wilford on snowdrops, Dr Tiziana Ulian on the value of nature, Lara Jewitt and Carlos Magdalena on how to grow the world's rarest plants.

Finally, to my wonderful family: Caroline, Tom and Oliver, whose curiosity is a daily motivation.

The Royal Botanic Gardens, Kew would like to thank the following people: Craig Brough, Julia Buckley, Kat Harrington, Anne Marshall, Lynn Parker and Kiri Ross-Jones of Kew's Library, Art and Archives; Paul Little for digitization; and Issy Wilkinson at Welbeck.

CONTENTS

ARAUCARIA IMBRICATA *Pav.*

♄ *Chili.* *Plein air.*

452.

INTRODUCTION

From the mud surrounding a hot Rwandan spring to a single mountainous ridge in the Philippines, plants have evolved to survive in extreme and remote corners of the Earth. Dependent on a constant environment, specific pollinators or a solitary method of seed distribution, these plants have a fragile relationship with thc world around them.

Our changing relationship with nature is making once abundant plants scarce and pushing naturally rare plants to extinction. Humanity's relationship with the natural world has shifted from stewardship and mutual benefit to exploitation, pushing land to its productive limit and causing irrevocable harm to biodiversity. Herbivorous animals introduced from distant lands to pristine habitats graze endemic plants to extinction, while pests and diseases transported by global trade infect, weaken and kill species with no immunity.

Climate breakdown is a constant pressure, pushing weakened and fragmented plant communities closer to the edge. Droughts, torrential rains and soaring temperatures are creating new environments where native plant species can no longer thrive. Plants introduced by human activities may become the new winners, swiftly colonizing disturbed habitats and outcompeting weakened indigenous species.

Richly illustrated with materials from the Library, Art and Archives at the Royal Botanic Gardens, Kew, this book explores what makes plants rare, celebrating the extraordinary powers of evolution that allow plants to exploit the planet's niches, no matter how remote or hostile. The threats ranged against our plants and the economic reasons driving biodiversity loss are explored through a global exploration, from Chile to the Chatham Islands, from the Himalayas to Herefordshire.

Among habitat loss and extinctions, there are solutions and hope for the future. Brilliant scientists, conservationists and horticulturists are discovering new species, working with local governments and communities to conserve biodiversity and developing new propagation methods to save rare plants.

It's a race against time, but advances in DNA sequencing, remote-sensing vast habitats and species distribution modelling are powerful new plant conservation tools. The Royal Botanic Gardens, Kew, works with partners around the world to stop biodiversity loss, through banking seeds of threatened plants, building evidence to conserve habitats and researching the value of wild plants: as crops, drugs and fibres of the future.

By understanding the value of nature and how it can benefit us if we protect it, we can shape a new approach to conservation. Some plants are inherently rare, while others have become rare through our actions. Understanding why these losses have taken place and the value of what we still have can stop biodiversity loss before it's too late.

Ed Ikin

OPPOSITE: *Araucaria araucana* (as *Araucaria imbricata*) from Louis Van Houtte, *Flore des serres et des jardin de l'Europe*, 1845.

RENALA

Adansonia grandidieri

The baobab (the genus *Adansonia*) has a highly dispersed wild distribution. *Adansonia* species are found in Australia, the Arabian Peninsula, southern Africa and Madagascar. Once thought to be a relic of the supercontinent Gondwana, recent research points to prehistoric humans as the agents of dispersal.

Six species of baobab are endemic to Madagascar. Sometimes called "the mother of the forest", these trees are a cornerstone of the island's ecology and are central to Madagascan culture and spiritual beliefs. The fortunes of Madagascar's baobabs vary, from relatively stable populations to species that are becoming increasingly scarce. Renala (*Adansonia grandidieri*), grows exclusively in the dry forests of south-west Madagascar, and its decline is cause for concern.

The otherworldly form (bulbous swollen trunk, diminutive root-like canopy) of a baobab follows remarkable function. Capable of persisting through extreme drought, this tree is adapted for survival on dry land. Due to a trait known as stem succulence, the trunk of *A. grandidieri* serves as a large-scale water-storage organ with a liquid capacity of over 1,000 litres. This contingency allows leaves to be produced during the dry season without any need for soil moisture or rain.

This species' ecology is no less remarkable than its physiology. Pollinated at night by lemurs and the straw-coloured fruit bat, the resultant fruit is valued for its nutrition. Rich in fats, calories and vitamin C, renala fruit is the most palatable. Although most of it is collected and used locally, ethical trading company Renala Naturals collects and cold presses seed from fallen fruit into a cosmetic oil for skin. Reportedly fairly traded at overseas export markets, this highly desirable product provides income that has the potential to support Madagascan conservation work, although this trade and its practices requires careful regulation.

The value and versatility of renala bark is starting to undermine the tree's survival. Ropes made from its bark are long-lasting and popular with cattle herders, builders and canoe makers. It also has a role in traditional Madagascan medicine (for treating calcium deficiencies) and this range of uses confers considerable economic value. The trade in

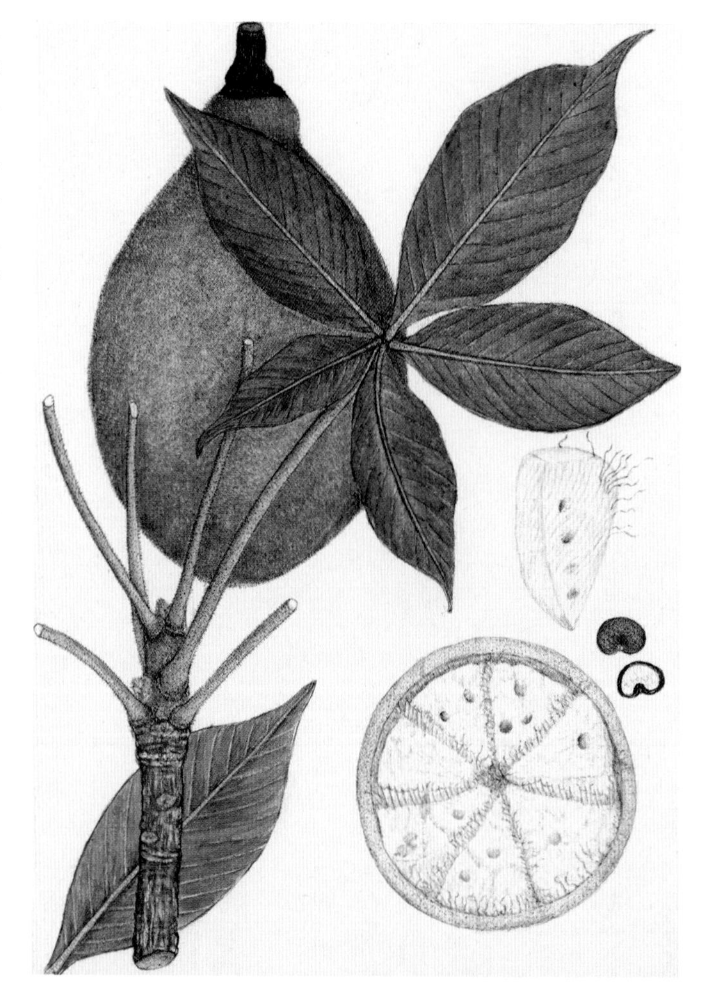

ABOVE: Original illustration of *Adansonia digitata* by Olive H. Coates Palgrave, produced for *Trees of Central Africa*, 1957.

OPPOSITE: Photograph of *Adansonia grandidieri* in Madagascar from Alfred Grandidier, *Histoire physique, naturelle et politique de Madagascar*, 1882.

its bark is tipping harvesting practices from sustainable to exploitative, as stripping reduces the vigour and resilience of specimens.

Though the tree is adapted to tough conditions, including tolerance of fire and limited habitat disturbance, there's growing evidence this meta-population in south-west Madagascar is losing viability. Land clearance for agriculture is accelerating, fires are increasingly frequent (undermining the regenerative capacity of young trees) and there's evidence of water pollution from the sugar industry. Baobabs are exceptionally long-lived trees, capable of growing for thousands of years. As a consequence, their reproductive strategy and subsequent capacity for regeneration is slow – a serious issue when combating threats that are growing fast.

Research on *A. grandidieri*'s future viability is timely: there are still about a million trees across its distribution, threats are clearly identified and sophisticated modelling has calculated the future decline of the species, providing vital foresight for proactive conservation. IUCN's Red List status of "Endangered" is as much a statement of future peril as current crisis, a clarion call to take action now.

Better regulation of local trade in bark products and further development of a fair-trade model for the desirable seed oil can reduce exploitation. Empowering local communities to acquire management rights for land with notable renala stands has been effective, and developing nurseries to grow young trees is showing promise. Can this conservation act as a model for other species? It's a proactive, multi-stranded approach based on evidence from advanced modelling – and it means acting now, before it's too late.

ABOVE LEFT: Original illustration of *Adansonia digitata* by Olive H. Coates Palgrave, produced for *Trees of Central Africa*, 1957.

ABOVE RIGHT: Herbarium specimen of *Adansonia grandidieri* collected in 1926, held at the Royal Botanic Gardens, Kew.

OPPOSITE TOP: *African Baobab Tree in the Princess's Garden at Tanjore, India*, by Marianne North, 1878.

OPPOSITE BOTTOM: *The Baobub*, by Thomas Baines, 1861.

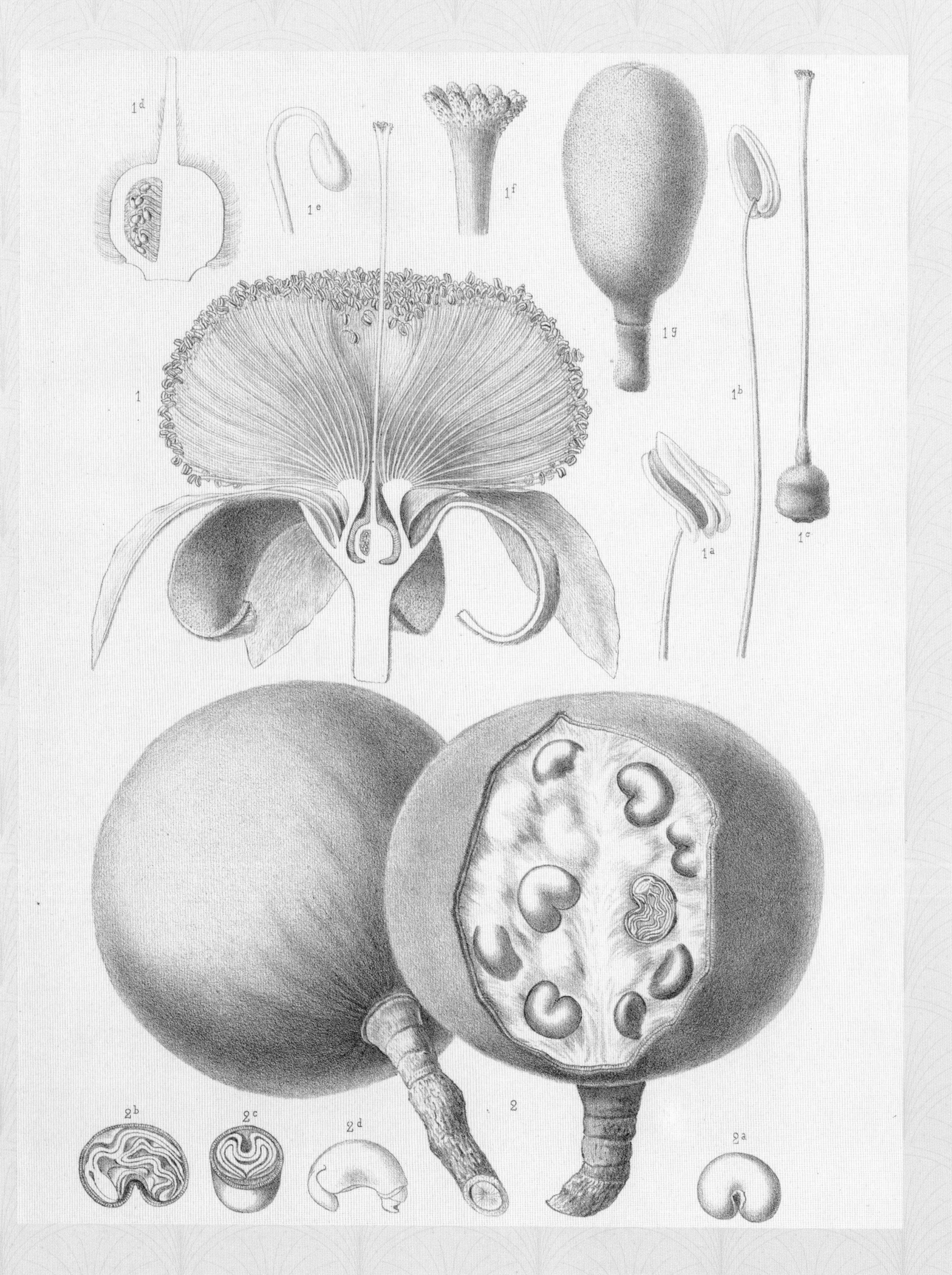
1d
1e
1f
1g
1
1b
1a
1c
2b
2c
2d
2
2a

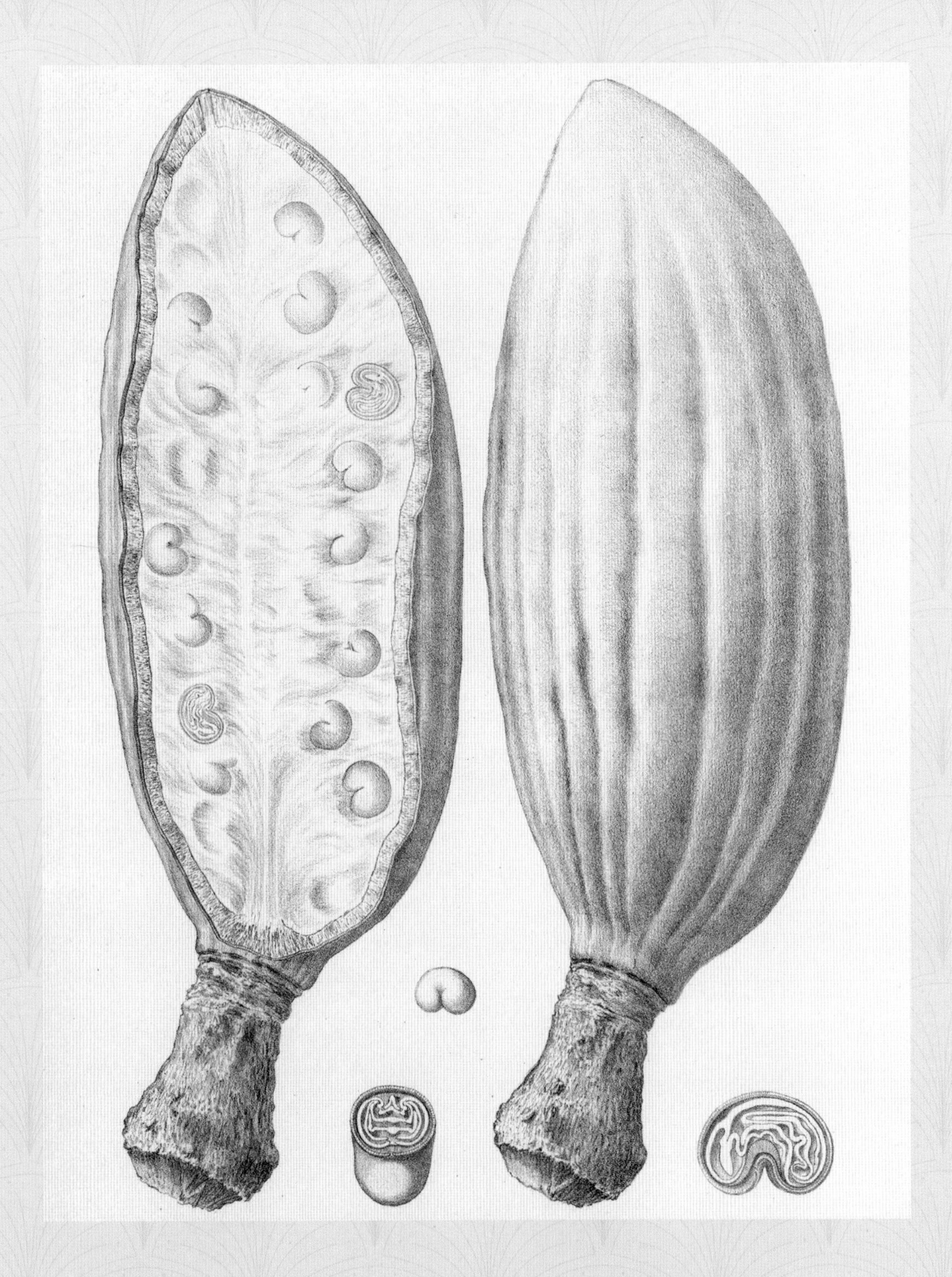

Drawings of *Adansonia* species by André Revillon d'Apreval in Alfred Grandidier's, *Histoire physique, naturelle et politique de Madagascar*, 1882.

This multi-volume work by French naturalist and explorer Alfred Grandidier (1836–1921) was the result of years of study on the island. Considered the first authority on the fauna and flora of Madagascar, Grandidier documented new species, with many named in his honour (including *Adansonia grandidieri*) and his observations of the landscape led to the creation of a map of the island, featured in this work and used by many subsequent explorers.

ALOE VERA

Aloe vera

Aloe vera is an enigma, a widely grown plant whose wild origins are a mystery, a puzzle slowly being unravelled by innovative research. There are few more ubiquitous plants: *A. vera* is cultivated globally, the familiar spiky foliage at home on sunny windowsills around the world. Gels, juices, yogurts, shampoos and moisturizers, all of which support a valuable global industry, are sold on the virtue of *A. vera*'s nurturing properties.

Aloe vera is a succulent, a broad term applied to plants evolved for drought. Specially adapted tissue stores water while an ingenious variation on photosynthesis allows pores to close during the day and then open at night, minimizing moisture loss. The mucilage that oozes from the plant's succulent tissue holds the secret to the plant's profitability: a product with a reputation for healing. Despite this renown, clinical evidence isn't wholly aligned with public sentiment.

A range of potentially beneficial plant chemicals have been identified in *A. vera*, but the clinical view of its ability to heal is still cautious. The anti-inflammatory effects of the plant on mice are cited and there is anecdotal evidence of healing burns and wounds, with the qualification that the mechanism is uncertain. A second *A. vera* product, a yellow sap called aloin that oozes from bruised leaves, is taken to clear the system but is regarded as toxic by healthcare agencies.

Aloe has a broad distribution in Africa, the Arabian Peninsula and islands across the Indian Ocean. Once part of the lily family, the 500 species in this genus are now classified in the Asphodelaceae, a family also including the red hot pokers (*Kniphofia*), a garden staple. The genus is filled with strikingly beautiful plants, with spectacular orange, red and yellow flower spikes rising from stiff succulent foliage. The long tubular flowers are adapted for insect and avian pollinators, with the South African sunbird a regular visitor.

OPPOSITE: *Aloe vera* (as *Aloe vera costa spinosa*) from Johann Wilhelm Weinmann, *Phytanthoza iconographia*, 1737.

RIGHT: *Aloe vera* (as *Aloe chinensis*) by Walter Hood Fitch, from *Curtis's Botanical Magazine*, 1877.

15-09

Ackd 8.1.09

Pharmaceutical Society of Great Britain,

17, BLOOMSBURY SQUARE, LONDON, W.C.

Jan 6th 1909

Museum Department.

ROYAL GARDENS 7 JAN 1909 KEW

The Director
Kew Gardens

Dear Sir

I have just had sent to me a leaf & a fruit spike from seeds of the aloe which yields the opaque aloes mentioned in Pharmacographia 2nd ed p. 686 as Natal aloes, & of which the botanical source was unknown. Ever since that date (1879) i.e. for 20 years I have been trying in season & out of season to get the plant & have only just succeeded. Some seeds were sent, & others seen quite fresh. I am sending you a few to grow. I compared my leaf & fruit spike at Kew Herbarium yesterday & so far as I can make out it is probably a new species that has been described under ~~the name~~ a varietal name under A. ferox. It seems to me that several species are confused under that name. The aloin yielded by this aloe is quite different from that of Aloe ferox. I showed Mr N. E. Brown my specimen & he is going I believe to look over the Aloe ferox group. The plant I have is or was apparently in cultivation at Sir Thos Hanbury's Garden. I am

Pharmaceutical Society of Great Britain,

17, BLOOMSBURY SQUARE, LONDON, W.C.

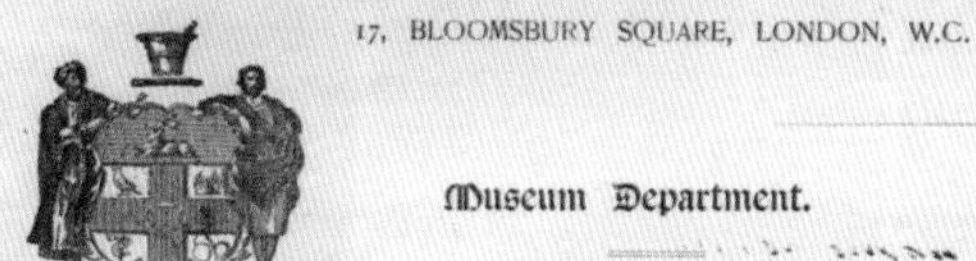

19

Museum Department.

writing to see if I can get a flowering specimen from there, as I want to publish the botanical source of the aloe after taking so much trouble in the matter. The seeds I enclose seem to be in good condition. I left a few with a capsule or two with Mr Brown for the Herbarium.

I am Dear Sir
Yours faithfully
EMHolmes

The plant so far as I could make out in the absence of flowers is the Aloe ferox var xanthostachys Berger & the Aloe supralaevis β Hanburii of the Flora capensis vol VI. p 327. But it is probably a distinct species. Baker's supralaevis is prickly on upper surface & the inflorescence is different to the true supralaevis.

Letter from Edward Morell Holmes, Curator of the Pharmaceutical Society of Great Britain, Bloomsbury Square, London, to Sir David Prain, Director of the Royal Botanic Gardens, Kew, 6 January 1909.

Holmes writes to Prain about a potential new species of *Aloe*, mentioned in the *Pharmaegraphia* as a Natal aloe of which the name is unknown. He believes this to be a species different from *Aloe ferox*, and runs through a number of species it could be. Holmes will try to aquire a flowering specimen, and has also enclosed some seeds for Prain to grow. Sir David Prain was Director of Kew from 1905–22. A trained physician with an interest in botany, Prain joined the Indian Medical Service and became Curator of the Herbarium and later Director, at the Royal Botanic Garden, Calcutta, until he took up the role at Kew in 1905.

OPPOSITE: *Aloe vera* (as *Aloe vulgaris*) by Pierre Joseph Redouté from Augustin Pyramus de Candolle, *Plantarum historia succulentarum*, 1799–1837.

Despite its prevalence, *A. vera* is not found in the wild and there's no record of its final act of extinction. Kew scientist Dr Olwen Grace is leading the detective work to discover its true origins. The starting point for unravelling the *A. vera* mystery is the wealth of cultivated plants supplying industry. By understanding how much or how little they vary and sequencing the DNA that lies within, connections can be formed between crop plant and wild relative. Sequencing *A. vera* DNA and comparing it to other members of the genus creates an extraordinary molecular family tree that is known as a phylogeny.

DNA sequencing has transformed our understanding of how plants relate to one another, with species being separated and joined, novel families created and evidence for new, as-yet-undiscovered species generated. The degree to which species are related is shown through branching diagrams that illustrate common ancestors and their distantly evolved relatives.

Aloe vulgaris.

OPPOSITE: *Aloe vera* (as *Aloe vulgaris*) from John Sibthorp, *Flora Graeca*, 1823.

ABOVE LEFT: *Aloe vera* (as *Aloe vera minor*) from Johann Wilhelm Weinmann, *Phytanthoza iconographia*, 1737.

ABOVE: *Aloe vera* (as *Aloe vulgaris*) from Theodor Friedrich Ludwig Nees von Esenbeck, *Plantae medicinales*, 1828.

The point in time when a new species becomes distinct and starts to radiate away from relatives can be clearly mapped. This could be the clue to placing *A. vera* in its original wild home with closely related species that are likely to share the same geographic region. Dr Grace's research places *A. vera* within the Arabian part of the genus's distribution, ending decades of speculation among taxonomists.

If *A. vera*'s precise wild home is discovered, should the plant be reintroduced there? This isn't as abstract a concept as it sounds. With the wild origins of *A. vera* lost, the benefits of this extraordinarily valuable plant cannot go to the rightful owner. Exclusive provenance can transform the value of a product – think "single-estate coffee" or "DOP Parmesan". The value of a hypothetical "wild-origin *Aloe*" could have transformative impact on the region selling it – not only bringing *A. vera* home, but also giving it renewed purpose.

3

MONKEY PUZZLE

Araucaria araucana

The national tree of Chile, the monkey puzzle found a new home far from its mountainous South American origins: on the lawns of Britain's stately homes. An imposing, reptilian presence that is capable of living 2,000 years, the ease with which this distant visitor adapted to garden conditions is testament to Britain's nurturing temperate climate. Success in British gardens contrasts strongly with challenges at home: the monkey puzzle is now endangered in its native Chile and Argentina due to uncontrolled wildfires, overgrazing and pressure from logging.

Monkey puzzles occupy the central mountainous regions of Chile and Argentina, growing at altitudes between 600 and 1,600 metres, in a climate of cool wet summers and cold winters. Its Chilean range is from Biobío to Valdivia, and in Argentina it occupies the province of Neuquén.

The monkey puzzle is adapted to harsh, rocky, even volcanic habitats, and colonizes ground quickly after eruptions have cleared a site. Tough scaly leaves – evolved to repel browsing dinosaurs – twist in spirals around strong branches, becoming spikier with age.

Trees show three distinctive modes of growth as they age. Young plants are proudly upright, with vigorous leaders extending up to a metre in wet summers. Middle age ushers in a complex pyramid form, with intricate layers of interwoven branches. Mature trees resemble a bristling umbrella – those spiralling branches are now 40 or 50 metres above the ground, with clean trunks liberated by years of dropping limbs (a common consequence of dry summers).

Gaze into the canopy of a lofty, mature monkey puzzle and you may spot enormous cones with a diameter of up to 20 centimetres. These cones resemble armoured coconuts and hold up to 200 seeds. Monkey puzzles are dioecious, with separate male and female trees.

Monkey puzzles have an exceptionally fine timber that is straight-grained and free of knots. As Chile developed its national infrastructure, monkey puzzles supplied robust, long-lasting railway sleepers. But these versatile trees also inspired fine carpentry – everything from pianos to skis. Seeds from their spiky cones are starchy, calorific and delicious. Made into a mash and fried or boiled by

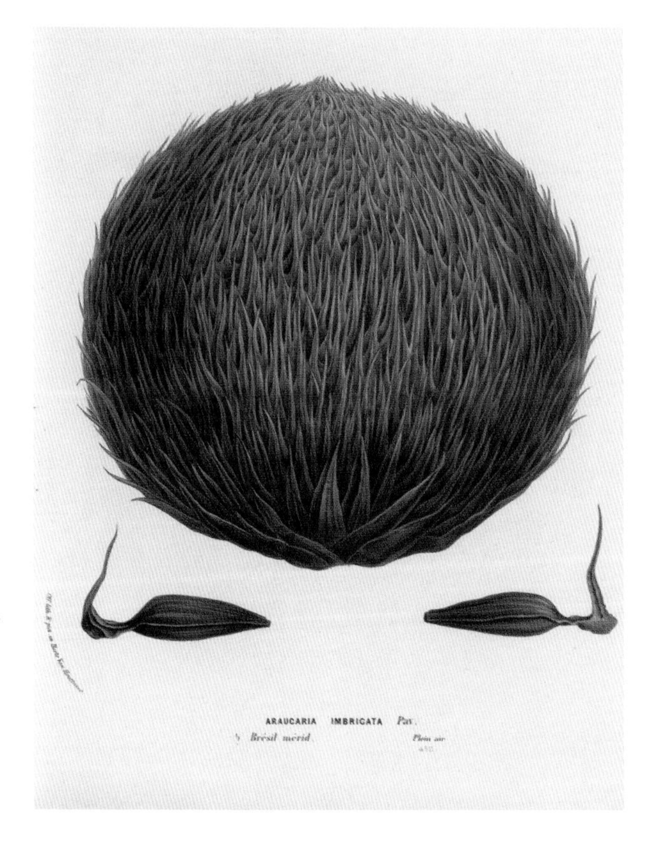

ABOVE: *Araucaria araucana* cone (as *Araucaria imbricata*) from Louis van Houtte, *Flore des serres et des jardin de l'Europe*, 1862–65.

OPPOSITE: *Araucaria araucana* (as *Araucaria imbricata*) from Louis van Houtte, *Flore des serres et des jardin de l'Europe*, 1845.

indigenous Chileans, the cones have a flavour that is often compared to sweet chestnut.

Across the ocean in Whitby, England – home of Dracula, goths and dramatic coastal views – tourist visits are often rounded off by the purchase of Whitby jet as a souvenir. Purchasers admiring their new jewellery may be surprised to learn the dense, shiny and light jet is fossilized monkey puzzle, connecting Yorkshire to its primitive past, when dinosaurs roamed the now-pastoral landscape.

Gardening creates new contexts for plants, sometimes taming them into mannered settings and other times encouraging wild naturalization. Transposing the monkey puzzle from forests on Andean slopes to single ornamental specimens diminishes some of this noble tree's rugged impact. Should your garden be capacious enough to hold several, they look best in large stands.

Araucaria araucana undoubtedly thrives best in cooler, wetter garden climates, with impressive plantings in the coastal English counties of Devon and Cornwall, in Ireland and on the west coast of Scotland. Drier climes force trees to drop an excess of lower limbs to reduce water transpiration, leading to ungainly "toothpick" specimens, which are not a fair representation of the species.

However, Chile's national tree is under grave pressure in its native habitat. Rated as "Endangered" on IUCN's Red List, monkey puzzles are felled for timber with subsequent regeneration limited by fire or grazing, fragmenting continuous forest populations. Monkey puzzles have adapted to combat fire with epicormic buds capable of post-burn sprouting, but the increasing frequency and intensity of fires, often deliberately started, overwhelms their ability to regenerate. Areas cleared of indigenous monkey puzzles are often replanted with fast-growing exotic tree species such as *Eucalyptus*. The slow regeneration of this tree, which evolved long before the Anthropocene, means it is now

OPPOSITE: *Araucaria araucana* (as *Araucaria imbricata*) from Louis van Houtte, *Flore des serres et des jardin de l'Europe*, 1877.

ABOVE: *Male cones of Araucaria araucana*, by Marianne North, *c.* 1880.

F. J. Drier See Regels Gartenflora 1865 p. 193 1901

BOTANIC GARDENS AND NURSERY.

E. TIDMARSH, Curator.

Grahamstown, Aug 19 1901

ROYAL GARDENS 19 SEP 01 KEW

Sir W Thiselton Dyer

Dear Sir

I this day post to you seeds of Araucaria Bidwillii in two different stages of their peculiar method of germination which is not noted in any work to which I have access, the germination of Rhizophora, as shown by Kerner & Oliver, is in some respects similar

The germination of Araucaria excelsa, is, so far as I now remember, much as others of the Conifers, A Brasiliana germinates after the way of Palms, Crinums &c, the root thrust into the soil and the plumule rising from some point above the surface of the soil, the seeds still containing the cotyledons remaining attached for a considerable time

Examination of the specimens now sent will show that the radicle having penetrated the soil, then develops into a woody, carrot like extension, the future base from which the tree is to proceed, the long root then separates at the top of the carrot-like process leaving free the point of the leading shoot, which after some weeks begins to extend, producing minute leaves

Of course persons who have grown these trees

BOTANIC GARDENS AND NURSERY.

E. TIDMARSH, Curator.

Grahamstown, 189

from the seeds must be acquainted with the fact as stated above; some twenty years or more back I had sent to me some of these "wooden Carrots" which had been recovered from a wreck, and from these we grew a number of fine trees, I should not be surprised to hear that the one now sent to Kew, was on arrival still in a condition to grow, this would be an interesting fact

I may further mention that our trees of A Bidwillii, for years produced female cones only, a year ago however, during an extreme drought the trees gave a crop of pollen bearing cones, the result being that we got a supply of fertile seed, or should we say fruit? If I remember rightly, some 30 years since, the Araucaria Bidwillii in the large Conservatory at Kew gave male cones on what had till then thought to be a female tree

I remain

Faithfully yours

E Tidmarsh

P.S. Should the particulars be in your opinion worthy of publication perhaps you would bring the fact to notice in the proper quarter ET

Letter from Edwin Tidmarsh, Curator of the Botanic Gardens, Grahamstown, to Sir William Thiselton-Dyer, Director of Kew, 19 August 1901.

Tidmarsh was Curator at the Botanic Gardens, Grahamstown, for over 30 years, and in this letter sends Thiselton-Dyer seeds of *Araucaria bidwillii* in different stages of germination, and describes the state of the seeds and their future development, including the type of cones produced by mature trees.

rapidly outpaced by its fast-changing environment.

Botanic gardens, in collaboration with Chilean partners, have a crucial role in the conservation of this species, collecting seeds from threatened wild stands and raising them in their living collections. A physiological quirk of the monkey puzzle makes another primary conservation tool – seed banking – moot. Monkey puzzles are recalcitrant, meaning their seeds cannot survive the process of desiccation and chilling (normally down to -20°C) that seed banks apply as standard procedure.

Growing monkey puzzles in living collections and planting them in our botanic gardens and arboretums are therefore the only effective banks of genetic material for future Chilean habitat restoration. A joint expedition between the Royal Botanic Gardens, Kew, Royal Botanic Garden Edinburgh and the Forestry Commission collected thousands of seeds from the last Chilean coastal distribution of the monkey puzzle in 2009. Trees raised from these seeds are now thriving in Kew's wild botanic garden, Wakehurst; Edinburgh's mountainside Benmore Botanic Garden; and the Forestry Commission's Bedgebury National Pinetum and Forest.

Large, genetically diverse stands of healthy young trees offer hope for this iconic species, a conservation resource 8,000 miles away that may one day return home.

ABOVE: *Seven Snowy Peaks seen from the Araucaria Forest, Chili* by Marianne North, 1880.

CHATHAM ISLAND CHRISTMAS TREE

Brachyglottis huntii

About 800 kilometres off the east coast of New Zealand – their next closest neighbours are Antarctica to the south and Chile yet further east – lie the Chatham Islands. Gentle in topography and climate and profoundly isolated from other land masses, they are home to exceptional biodiversity, with notable endemism and gigantism.

The Chatham Islands host a rich array of bird species, and are utilized by two species of albatross as a southern-Atlantic staging post. Some 18 different bird species occur only on these 900 square kilometres. Warm subtropical currents from the north meet the icier sub-Antarctic southerly currents to create conditions for copious sea life. As a result, the Chatham Islands host some of the world's richest fishing waters.

Formerly part of the ancient supercontinent Gondwana, and long separated from New Zealand, the Chatham Islands were augmented by a volcanic eruption – and they now boast flora that is equally distinctive. Equitable climate, peaty soil and relatively late arrival of European settlers stimulated a distinctive suite of plants, including extraordinary endemics. Over 300 species, of which 47 are endemic, populate this modest archipelago, including the remarkable Chatham Island Christmas tree (*Brachyglottis huntii).*

Common names can be deceptive. Shared understanding of a plant or animal makes a common name like "daisy" or "robin" viable with no risk of confusion with another species. Applied internationally, common names become problematic. Terms like "pine" are applied to species from varying taxonomic backgrounds and so the risk of confusion grows, making scientific names the most desirable shared terminology. The Chatham Island Christmas tree bears no relation to the *Abies* (fir) or *Picea* (spruce) species cut in December to furnish festive homes.

OPPOSITE: *Brachyglottis huntii* from Georgina Burne Hetley, *The native flowers of New Zealand*, 1888.

RIGHT: Herbarium specimen of *Brachyglottis huntii* collected by Georgina Burne Hetley in 1887, held at the Royal Botanic Gardens, Kew.

23

Differences, such as are pointed out as existing between these two Chathamian plants, may be noted in several Australian species.

SENECIO HUNTII.

Arboreous, clammy; branchlets and peduncles glandulous-downy; *leaves lanceolate, entire, firm, reflexed at the margin, from subtle downs pale beneath, with unenlarged base sessile; above gradually glabrescent;* their veins immersed; *panicle compact, terminal,* surrounded by leaves, somewhat pyramidal, producing numerous flower-heads; peduncles copiously beset with very short brown gland-bearing hair; *capitula ligulate, with very numerous flowers;* involucre from semiovate verging to semiglobose, supported by few linear-subulate bracts; its scales about 13, unequal, mostly blunt, not much shorter than the discal flowers, somewhat glandular-downy; alveoles of the receptacle toothless; ligules of the female flowers not much longer than their tube, entire; anthers almost entirely exserted; bristles of the pappus nearly as long as the discal corollæ, twice or thrice as long as the glabrous achenia, not thickened towards the apex, almost biseriate.

On damp localities of woods growing generally in patches; rare in Chatham-Island, common in Pitt-Island.

A tree, often attaining according to Mr. Travers a height of 25′, called "Rautine" by the aborigines. Branchlets cicatricose, seemingly soon defoliated. Leaves, as far as seen, about 3″ long, ⅔–1″ broad, with spreading primary and closely netted immersed secondary veins. Panicle devoid of any long universal peduncle, interspersed with some generally short and narrow floral leaves. Ultimate peduncles somewhat or hardly longer than the capitula. Bracts near the apex of the special peduncles and around the involucres 1–2‴ long. Scales of involucre 2½–3‴ long, linear- and lanceolate-oblong, bearded at summit, imperfectly downy at the back, finally separating from each other. Ray-flowers 15–18; their ligule 2–3‴ long, lanceolate-oblong, yellow; their style partially exserted. Disk-flowers about 40; their corolla barely 3‴ long, campanulate above the middle, five-toothed at the summit. Anthers about 1‴ long. Stigmata exserted. Achenia measuring about 1‴ in length, furrowed, slender. Pappus white, very tender, composed of 50–60 serrulate indistinctly biseriate bristles.

This plant received its specific signification in order that the name of Mr. Frederick Hunt may also phytologically for ever be identified with that of the small isle, of which he was the first and is

24

still the principal European occupant, and in which this remarkable species forms such a prominent feature of the primeval vegetation. Mr. Hunt is moreover highly entitled to this mark of respect for the kind assistance which he afforded to the young traveller in his exertions of rendering known the vegetable products of these islands. A hope is simultaneously expressed, that as a permanent resident there the hospitable settler of Pitt-Island may also hereafter advance our cognizance of the vegetation around him.

In its arborescent growth Senecio Huntii has probably amongst the many hundreds of its congeners in the Victorian and Tasmanian Senecio Bedfordii (Bedfordia salicina, Cand. Prodr. vi. 441) and in the New Zealand S. Forsteri (J. Hook. Fl. Nov. Zeel. i. 148, t. xl.) its only rivals. In deep humid forest-gullies, favorable for its luxuriant development, S. Bedfordii equals S. Huntii in height, and thus both excel as the tallest all others of a cosmopolitan genus, which is recognized next to Solanum as the richest of all in species.

In the systematic series S. Huntii may perhaps find its place nearest to the exclusively Tasmanian S. Brownii (Centropappus Brunonis, J. Hook. in Lond. Journ. of Bot. vi. 124; Flor. Tasman. i. 225, tab. lxv.), which although much smaller exceeds also most other species in size. Both plants have many characters in common, but the Tasmanian plant differs in its smoothness, narrower flat leaves, smaller panicles with less large capitula and often shorter peduncles, much less numerous flowers of the heads, fewer and shorter scales and broader and shorter bracts of the involucre as also more grossly serrated upwards thickened bristles of the pappus.

S. Huntii approaches likewise in many characters to S. glastifolius (J. Hook. Fl. Nov. Zeel. i. 147, tab. xxxix.), which recedes again in its smoothness, in petioled often toothed leaves, in a less compact inflorescence with fewer and larger capitula, in larger bracts and also in upwards somewhat thickened pappus-bristles.

The transit of Bedfordia and Centropappus to Senecio, indicated by Dr. J. Hooker, is rendered sufficiently clear by the discovery of this Chathamian species.

SENECIO RADIOLATUS.

Upper leaves above the middle toothed and often also pinnatifid, below the middle entire, with cordate or auriculate base sessile, beneath somewhat arachnoid-downy; their lobes almost semilanceolate; flower-heads numerous, corymbose-paniculate; *receptacle pitted with toothed alveoles;* involucres from cylindrical gradually some-

Also known by its indigenous Moriori name of "rautini", the Chatham Island Christmas tree is a towering daisy in the genus *Brachyglottis*, with specimens growing over 8 metres tall. Trunk-like structures (a trait known as caulescence) support a spreading canopy of long spear-shaped leaves covered in fine, downy silver hairs. Topping this mop of foliage is a surreal sight for those who perceive daisies as diminutive lawn-dwellers: a vibrant veneer of sunny yellow ragwort-like flowers.

B. huntii, along with another tree daisy, *Olearia traversiorum*, formed the climax vegetation of the islands, sheltering other remarkable endemics such as the Chatham Island forget-me-not (*Myosotidium hortensia*). From 1840, European settlers arrived in small numbers, and intensive agriculture emerged as a significant use of the land after the First World War. With agriculture came modern cattle breeds that grazed crops and native plants, quickly disrupting flora and fauna unused to disturbance.

Changing landscape, introduced plants (including European gorse), and bracken rampantly colonizing disturbed ground rapidly altered the Chatham Islands' vegetation in the twentieth century, undermining dependent native fauna. Plant diseases novel to islands, including *Phytophthora* and Verticillium wilt, found an unwitting and receptive new host in the Chatham Island Christmas tree, the plant incapable of self-defence.

The Galápagos Islands have made their biodiversity a financial asset, using a tourist economy to conserve their extraordinary flora and fauna in a stable environment. The Chatham Islands and many other biodiverse archipelagos have not been so fortunate, with competing economic needs unsettling a delicate ecological equilibrium. Can a balance between supporting human livelihoods and biodiversity be found in these far-flung locations?

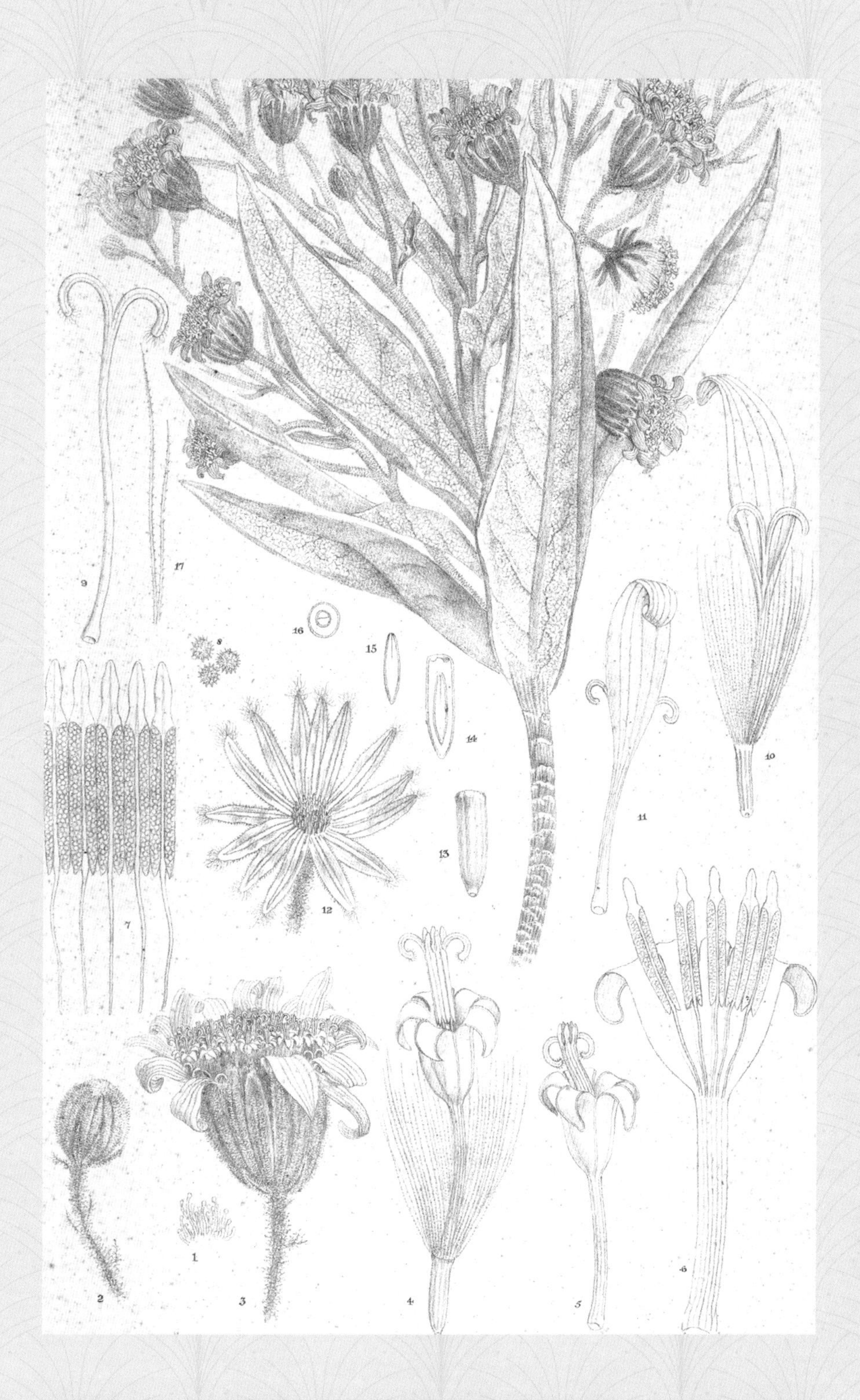

Pages from Ferdinand von Mueller, *The Vegetation of the Chatham Islands*, 1864.

The text (on page 30) features von Mueller's description of *Brachyglottis huntii* as *Senecio huntii*, with the above accompanying illustration detailing the various plant parts. This 86-page volume was published during von Mueller's time as Director at the Royal Botanic Gardens, Melbourne. Von Mueller is famous for naming many Australian plants, as well as founding the National Herbarium of Victoria.

LEFT: The Chatham Island forget-me-not, *Myosotidium hortensia* (as *Cynoglossum nobile*) by Walter Hood Fitch from *Curtis's Botanical Magazine*, 1859.

OPPOSITE: *Brachyglottis monroi* (as *Senecio monroi*) by Matilda Smith from *Curtis's Botanical Magazine*, 1917.

8698
1
2
3
4
5
6
7

ANGEL'S TRUMPETS

Brugmansia

The family Solanaceae is intertwined with humanity's successes and failures. The cultivation of the South American species *Solanum tuberosum*, better known as the potato, brought economic prosperity to Europe, and then economic disaster to Ireland. The dried leaves of *Nicotiana tabacum* offer vast wealth to those who sell them, and dependence, illness or worse to those who smoke them.

Solanaceae is filled with plants that nourish and plants that kill. Aubergines and peppers are globally consumed crops, while few plants are more lethal than *Atropa bella-donna*. Another member of the Solanaceae contains psychoactive chemicals so potent it holds mythical status in its native South America: the genus *Brugmansia*.

Brugmansia is a genus of just seven species, native to South America, with a distribution across Brazil, Venezuela and into Chile. This distribution is academic, as the genus is extinct in the wild even as it paradoxically flourishes in cultivation. There are South American stories of lore about bizarre dreams caused by sleeping under *Brugmansia* flowers and of unconsciousness induced by merely inhaling their fragrance. This hallucinogenic power is widely documented, and *Brugmansia* has a vast range of uses for the indigenous peoples of South America.

Knowing the plants' psychoactive potency, it's perhaps too easy to ascribe sinister qualities to its appearance. What sinister threat lurks within those tightly furled flowers, their pendulous nature deterring inspection? In reality, the source of *Brugmansia*'s narcotic power is equally an ornamental asset. Cultivated forms of *Brugmansia* express a vibrant and occasionally distasteful spectrum of pinks, purples, yellows and whites, tones you'll find in any botanic garden with a frost-free climate or conservatory.

In the absence of frost, *Brugmansia* thrives – and there's a curious tension between the masses of people that experience angel's trumpets in cultivation and the mind-altering potential of the flowers. Many countries view cultivated *Brugmansia* as an invasive weed; they curse its ability to quickly colonize disturbed ground and the persistence of its woody stems in the face of removal.

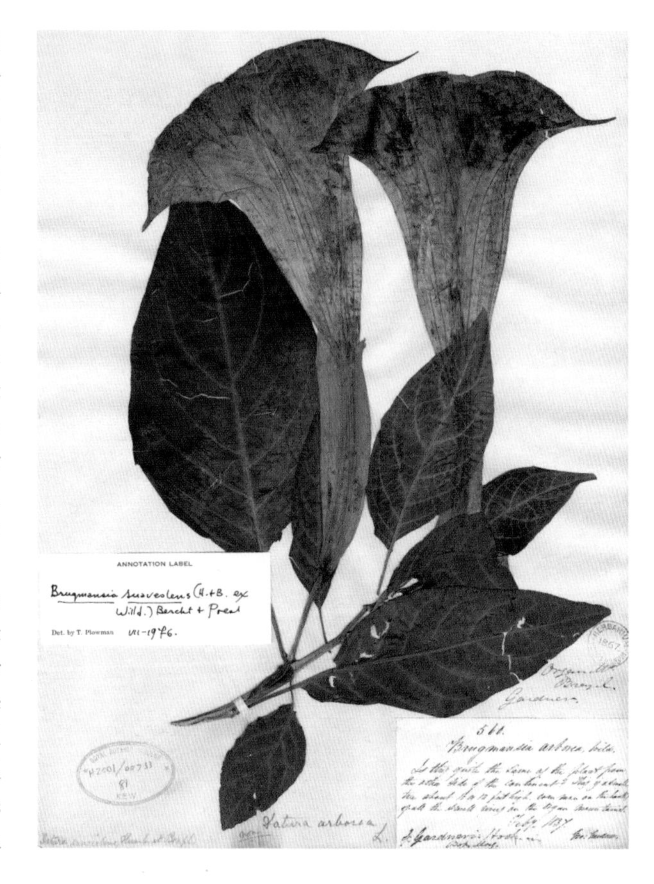

ABOVE: Herbarium specimen of *Brugmansia suaveolens* collected by G. Gardner, held at the Royal Botanic Gardens, Kew.

OPPOSITE: *Flowers of Datura and Humming Birds, Brazil*, by Marianne North, 1872. Commonly known as devil's trumpets, the *Datura* genus is closely related to *Brugmansia*.

LEFT: *Valley behind the Artist's House at Gordontown, Jamaica*, by Marianne North, 1872. *Brugmansia arborea* is the plant in the foreground, on the right-hand side.

Brugmansia flowers and fruits contain a range of alkaloid toxins, including scopolamine and atropine, a chemical suite shared with other "toxic" Solanaceae members such as deadly nightshade and henbane. Acting on the autonomous nervous system, these chemicals are capable of severe damage to the heart, eyes and digestive organs – and the volume of alkaloid toxins in mature flowers is potentially fatal. Children merely touching the flowers may suffer serious reactions. Yet, in a clinical environment, carefully controlled doses have positive effects: scopolamine has been applied to treat depression, while atropine aids in the treatment of Parkinson's disease.

The plant is central to several indigenous belief systems in South America, enabling divination, underpinning initiation ceremonies and warranting judicious application in medicine. Known as "mishas" in northern Peru, the plants' hallucinogenic and therapeutic powers are profoundly understood and respected. A strict code governing the preparation, application and use of *Brugmansia* products has evolved to utilize its powerful potential, contradicting clichés of "folk" medicine being uninformed or imprecise. The wisdom and inherent knowledge surrounding indigenous use of *Brugmansia* contrasts with the distrust it provokes in urban communities. Feared and reviled as toxic plants, their removal from gardens and cultivated land in South America may be eradicating the last "wild" genetic material left.

Extinct but invasive, lethal but medicinal, revered and feared, *Brugmansia* comprises plants of contradiction – at once beautiful, deadly, and vital.

CHAPTER VII

BORNEO AND JAVA

AFTER a fortnight at Government House, Sir William wrote me letters to the Rajah and Rani of Sarawak, and I went on board the little steamer which goes there every week from Singapore. After a couple of pleasant days with good old Captain Kirk, we steamed up the broad river to Kuching, the capital, for some four hours through low country, with nipa, areca, and cocoa-nut palms, as well as mangroves and other swampy plants bordering the water's edge. At the mouth of the river are some high rocks and apparent mountain-tops isolated above the jungle level, covered entirely by forests of large trees. The last mile of the river has higher banks. A large population lives in wooden houses raised on stilts, almost hidden in trees of the most luxuriant and exquisite forms of foliage. The water was alive with boats, and so deep in its mid-channel that a man-of-war could anchor close to the house of the Rajah even at low tide, which rose and fell thirty feet at that part. On the left bank of the river was the long street of Chinese houses with the Malay huts behind, which formed the town of Kuching, many of whose houses are ornamented richly on the outside with curious devices made in porcelain and tiles. On the right bank a flight of steps led up to the terrace and lovely garden in which the palace of the Rajah had been placed (the original hero, Sir James Brooke, had lived in what was now the cowhouse). I sent in my letter, and the Secretary soon came on board and fetched me on shore, where I was most kindly welcomed by the Rani, a very handsome English lady, and put in a most luxurious room, from which I could escape by a back staircase into the lovely garden whenever I felt in the humour or wanted flowers.

The Rajah, who had gone up one of the rivers in his gunboat yacht, did not come back for ten days, and his wife was not sorry to have the rare chance of a countrywoman to talk to. She had lost three fine children on a homeward voyage from drinking a tin of poisoned milk, but one small tyrant of eighteen months remained, who was amusing to watch at his games, and in his despotism over a small Chinese boy in a pigtail, and his pretty little Malay ayah. The Rajah was a shy quiet man, with much determination of character. He was entirely respected by all sorts of people, and his word (when it did come) was law, always just and well chosen. A fine mastiff dog he had been very fond of, bit a Malay one day. The man being a Mahomedan, thought it an unclean animal, so the Rajah had it tried and shot on the public place by soldiers with as much ceremony as if it had been a political conspirator, and never kept any more dogs. He did not wish to hurt his people's prejudices, he said, for the mere selfish pleasure of possessing a pet.

He had one hundred soldiers, a band which played every night when we dined (on the other side of the river), and about twenty young men from Cornwall and Devonshire called "The Officers," who bore different grand titles,—H. Highness, Treasurer, Postmaster-General, etc.,—and who used to come up every Tuesday to play at croquet before the house. Some of them lived far away at different out-stations on the various rivers, and had terribly lonely lives, seldom seeing any civilised person to speak to, but settling disputes among strange tribes of Dyaks, Chinese, and Malay settlers.

The Rajah coined copper coins, and printed postage-stamps with his portrait on them. The house was most comfortable,

Opening pages of Chapter 7 from Marianne North, *Recollections of a Happy Life*, 1892.

Marianne North was a remarkable Victorian artist and explorer, traversing the globe to record the world's flora. From 1871 to 1885 she visited America, Canada, Jamaica, Brazil, Tenerife, Japan, Singapore, Sarawak, Java, Sri Lanka, India, Australia, New Zealand, South Africa, the Seychelles and Chile. She was well connected and often stayed with people of influence and standing, such as the Rajah and Rani of Sarawak, mentioned in this extract.

The Marianne North Gallery at Kew houses 832 of North's paintings; the layout of which was instructed by North and remains as such today.

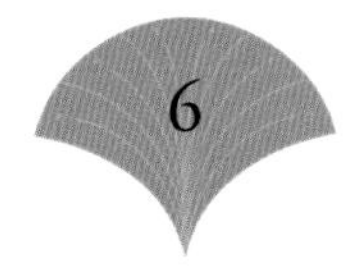

SEA MARIGOLD

Calendula suffruticosa subsp. *maritima*

Coastal plants exist in a dynamic growing environment. Storms, landslips and salt spray make unstable, hostile habitats and only plants adapted to high levels of disturbance (sometimes called ruderals) thrive. The plants that do survive on cliffs and dunes offer a range of desirable services, stabilizing shifting land and providing food and nesting material to resident fauna. Natural forces already make this environment unpredictable and hostile, and adding human activity into the coastal equation is even more destabilizing.

Calendula suffruticosa subsp. *maritima* is an extremely rare representative of an otherwise abundant plant, a coastal exponent of the sunny optimism embodied in the genus *Calendula*. Capable of covering clifftops and shorelines with expansive drifts of compact golden flowers, the sea marigold is only found in the Trapani region of western Sicily and the scattered islets of nearby Formica and Favignana.

Geographic isolation made *C. suffruticosa* subsp. *maritima* a distinct species, and its distribution is inherently limited to Sicily, but this small population is growing smaller at an alarming rate. The plant's total wild territory is a mere 10 square kilometres, so it's not surprising that IUCN classify it as "Critically Endangered" on its Red List.

Being exposed to the winter vagaries of the western Mediterranean is a potent selection pressure, but this is not *C. suffruticosa* subsp. *maritima*'s only threat. Increasing coastal development, in the form of tourist resorts and harbours, are converting coastal habitat to human residences, while invasive plant species encroach.

Introduced plants like the South African *Carpobrotus edulis* (Hottentot fig) delight in Mediterranean coastal conditions, forming dense impenetrable carpets and entrancing unenlightened tourists with their thick succulent leaves and vivid pink flowers. Indigenous plants cannot seed or spread in the presence of such a monoculture. Roadside verge cutting is a further pressure on this imperilled population.

The sea marigold also faces a more familiar threat: another *Calendula* species. DNA-based phylogenetic trees have established *C. suffruticosa* subsp. *maritima* as a genetically distinct species, but it's still closely related enough to common *Calendula suffruticosa* subspecies *fulgida* to hybridize with it.

OPPOSITE: Original illustration of *Dimorphotheca chrysanthemifolia* (as *Calendula chrysanthemifolia*) from *Curtis's Botanical Magazine*, 1891. *Dimorphotheca* and *Calendula* are closely related genera in the Asteraceae family.

ABOVE: Herbarium specimen of *Calendula suffruticosa* collected in 1846, held at the Royal Botanic Gardens, Kew.

The crossed new seedling is no longer *C. suffruticosa* subsp. *maritima*, reducing the reproductive chances for a short-lived species, and making its hybrid offspring a potential competitor.

Calendula species are a globally valuable crop as a popular herbal remedy. Used to treat burns, wounds and inflammation, *Calendula*-based gels, oils and creams are sold widely. There's a solid phytochemical basis to this healing reputation – the plants contain tripertinoids, a group of compounds proven to reduce inflammation isolated from plant tissue. Despite this, clinical evidence for these benefits is still limited.

The threat of *C. suffruticosa* subsp. *maritima*'s imminent extinction has focused conservation efforts, as local authorities are galvanized by its "Critically Endangered" IUCN rating. More effectively enforced reserves, a concerted Hottentot fig eradication scheme and a conservation programme collecting, banking and propagating pure *C. suffruticosa* subsp. *maritima* offer future hope, supported by a €1 million investment from the European Commission.

Placing a greater value on *C. suffruticosa* subsp. *maritima*'s ecosystem services – stabilization of highly friable coastal soils, nectar for pollinators and recycling of the rotten sea grass it establishes on – could shift perceptions of this highly threatened, too-rare plant. Until then, the predictable battle of commerce versus conservation rages, with a beautiful, fragile plant caught in the middle.

ABOVE: Illustration from Johann Hogenberg, *Series of Prints with Flowers and Animals in a Landscape*, 1600–05. Calendula can be seen on the left of this illustration. Rijksmuseum.

LEFT: Original illustration of *Dimorphotheca tragus* as *Calendula tragus* from *Curtis's Botanical Magazine*, 1818.

OPPOSITE: *Calendula officinalis* from Hermann Adolf Köhler, *Medizinal-Pflanzen*, 1887.

4
4 a
3
2
1
6
5
A
7
8
9
10

LOBSTER CLAW

Clianthus puniceus

The extrovert appearance of *Clianthus puniceus* beguiles and surprises. Its common names, lobster claw or parrot's beak, reflect a flower hard to comprehend as plant in origin. Dense clusters of scarlet claw-shaped flowers emerge in profusion during late spring, an exotic sight in frost-free gardens and glasshouses and an increasingly rare one in its native New Zealand. Abundant in cultivation and scarce in wild distribution, *C. puniceus* highlights a paradox found in many garden plants.

Known by the Māori as "kowhai ngutukākā" or "kākābeak", *C. puniceus* appears to have been favoured by traditional land management, and use of the flowers as ephemeral jewellery has been recorded. Although those luxuriant flowers produce decent volumes of seed, held in pendulous pods (it's a member of the pea family, Fabaceae), the plant's distribution mechanism appears ineffective. *C. puniceus* and *Clianthus magnificus*, the only two plants in their genus, have the ability to "layer", wherein branches touching the ground form roots and become independent plants. However, this clones existing plants rather than generating new genetic variants from seed production.

A naturally low capacity for reproduction makes *C. puniceus* vulnerable to additional threats, and these came once New Zealand's human population started to change their use of the land. Primary forest clearance, invasive weeds and grazing from introduced herbivores, including pigs, deer, possums and slugs, dramatically increased the plant's rarity.

The scarcity of *C. puniceus* has impacted the New Zealand fauna with which it co-evolved. The bellbird feeds on its nectar, while an indigenous mite, *Aceria clianthi,* feeds on nothing else. Inevitably, this mite has become as endangered as its host plant. IUCN rate *C. puniceus* as "Endangered" on their Red List, with the wild population totalling little more than 150 plants in the east of New Zealand's North Island.

Debate persists about whether the native population is actually wild – some believe that a part of the current population is actually historic Māori plantings that were established close to settlements, a reflection of the plant's cultural value. In addition to its flowers being worn,

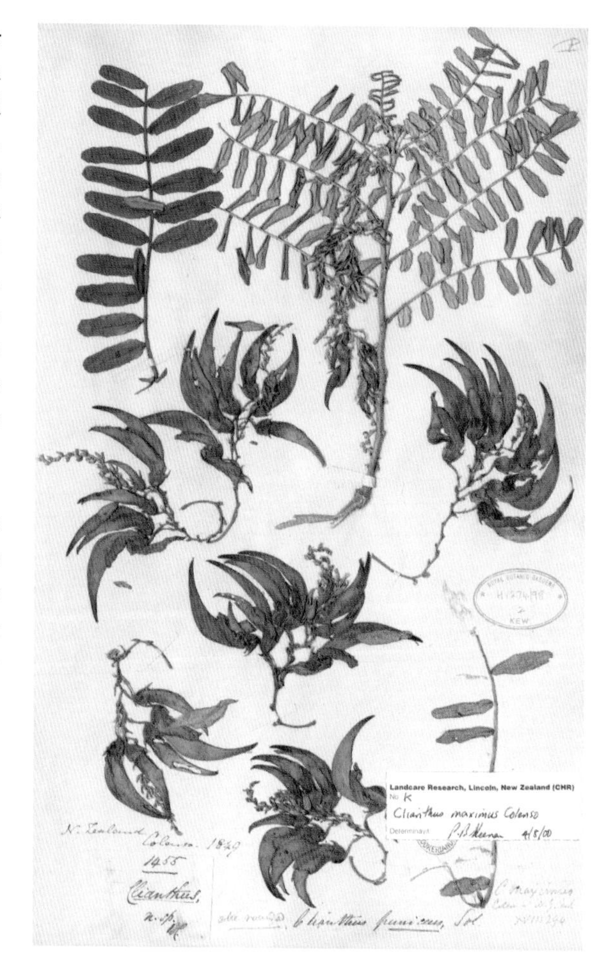

ABOVE: Herbarium specimen of *Clianthus puniceus* collected by W. Colenso in New Zealand in 1849, held at the Royal Botanic Gardens, Kew.

OPPOSITE: *Clianthus puniceus* (as *Donia punicea*) by Sarah Drake from *Edwards's Botanical Register*, 1836.

Clianthus puniceus.

Miss Taylor del.

CLIAN'THUS PUNI'CEUS.

CRIMSON CLIANTHUS, OR GLORY-PEA.

EXOGENÆ, OR DICOTYLEDONEÆ.

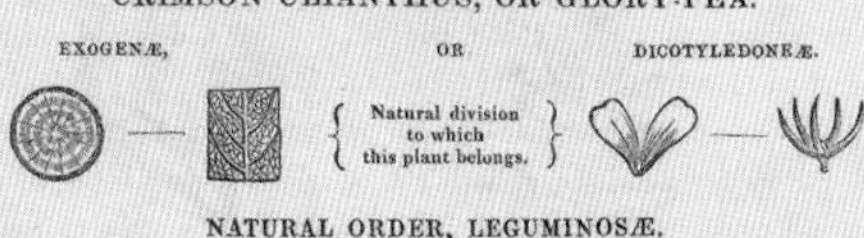

NATURAL ORDER, LEGUMINOSÆ.

CALYCIFLORÆ, OF DECANDOLLE.

Artificial divisions to which this Plant belongs

DIADELPHIA, DECANDRIA, OF LINNEUS.

No. 44.

GENUS. CLIANTHUS. *SOLANDER.* CALYX latè campanulatus, subæqualis, 5-dentatus. VEXILLUM acuminatum, reflexum, alis parallelis longius; carina scaphiformis, vexillo alisque multo longior, omninò monopetala. STAMINA manifestè perigyna, diadelpha, omnia fertilia. STYLUS staminibus duplo longior, versus apicem hinc leviter barbatus, stigmate simplicissimo. LEGUMEN pedicellatum, coriaceum, acuminatum, ventricosum, polyspermum, intus lanulosum, suturâ dorsali rectâ ventrali convexâ. SEMINA reniformia, funiculis longiusculis affixa. *LINDLEY* in Botanical Register, folio 1775.

SPECIES. CLIANTHUS PUNICEUS. *SOLANDER.* SUFFRUTICOSUS diffusus glaber, foliolis alternis oblongis subemarginatis, racemis pendulis multifloris, calyce 5-dentato, legumine glabro. *LINDLEY.* Botanical Register, 1775.

CHARACTER OF THE GENUS, CLIANTHUS. CALYX widely campanulate, nearly equal, 5-toothed. STANDARD acuminate, reflexed, longer than the parallel wings, keel skiff-shaped, much longer than the standard and wings, completely monopetalous. STAMENS manifestly perigynous, diadelphous, all fertile. STYLE twice as long as the stamens, towards the apex slightly bearded, stigma quite simple. LEGUMEN pedicellate, coriaceous, acuminate, ventricose, many-seeded, somewhat woolly within, dorsal suture straight, ventral suture convex. SEEDS kidney-shaped, attached by rather long chords.

DESCRIPTION OF THE SPECIES, CLIANTHUS PUNICEUS. STEM branched from 2-4 feet high, round, smooth, except when cracked, devoid of all pubescence save on the under surface of the young leaves, and on the green parts of the flower; branches green. LEAVES alternate, stipulate, oddly pinnate, of 8 pairs of folioles; folioles oblong, obtuse, subemarginate, distinctly alternate: stipules green, ovate, reflexed, very much smaller than the folioles. RACEMES pendulous, many-flowered; axis flexuous; BRACTS ovate, reflexed, very much shorter than the slender bracteolated pedicels. CALYX 5-toothed, teeth acuminate. STANDARD ovato-lanceolate, acuminate, reflexed, 2 inches long, externally of a rose-colour, internally of a deep blood-colour except when towards the base it is marked with interrupted white streaks or lines. WINGS of a blood-red colour, obtuse, about 1½ inch in length. KEEL quite monopetalous, acuminate, nearly 3 inches long, of a redish orange colour, pale towards the base. POD nearly 3 inches in length, dark-brown, veined. SEEDS kidney-shaped, brown, speckled with black spots.

REFERENCE TO THE DISSECTIONS.
1, Stamens and Pistil. 2, Calyx. 3, Wing. 4, Keel.

POPULAR AND GEOGRAPHICAL NOTICE. The enterprising naturalists, Banks and Solander, who accompanied Captain Cook, in 1769, first discovered this plant in the northern interior, of New Zealand; it was again discovered by the missionaries in 1831. Its native name is KOWAINQUTUKAKA or Parrot's-bill: but it is most justly entitled to the name, given by Solander, of Flower of Glory. A group of such shrubs would realize the description by the poet—

Of flowers that with one scarlet gleam
Cover a hundred miles, and seem
To set the hills on fire!

INTRODUCTION; WHERE GROWN; CULTURE. Mr. Richard Davis, Missionary Catechist at New Zealand sent the seed of Clianthus puniceus to the Rev. John Noble Colman, of Ryde, Isle of Wight, who sowed it as soon as it was received in the autumn of 1831. In the following spring they produced several fine plants. The specimen from which our drawing was made flowered in May, 1836, in the rich collection of William Leaf, Esq. Parkhill, Streatham. Cuttings strike root most readily under a hand-glass, indeed where its branches touch the ground, they will take root like Verbena Melindris. Trained to a southern wall, it will grow luxuriently, but notwithstanding its apparent health, during winter, in such situation, when spring succeeds, it betrays its southern origin, and either dies, or recovers with difficulty.

DERIVATION OF THE NAMES.

CLIANTHUS, from κλἕιος glory, and ανθος a flower. PUNICEUS, scarlet, from Punicus, of or belonging to Phœnicia, of which Tyre was famous for its dye of purple, said to be obtained from a species of shell-fish of the genus Murex.

SYNONYMES.

CLIANTHUS PUNICEUS, Solander, Manuscript in British Museum. Allan Cunningham in Transactions of Horticultural Society, New Series, Vol. I, p. 521, t. 22. Hooker in Botanical Magazine, folio 3584.

DONIA PUNICEA. George Don's General Dictionary of Gardening and Botany, Vol. II, p. 468.

Pages from Benjamin Maund and John Stevens Henslow, *The Botanist*, 1837–42.

Detailed description of *Clianthus puniceus* and accompanying illustration (see opposite).

Great friend and teacher to Charles Darwin, John Stevens Henslow collaborated with pharmacist and botanist Benjamin Maund on their five-volume journal *The Botanist*, published over a period of five years with hand-coloured engravings.

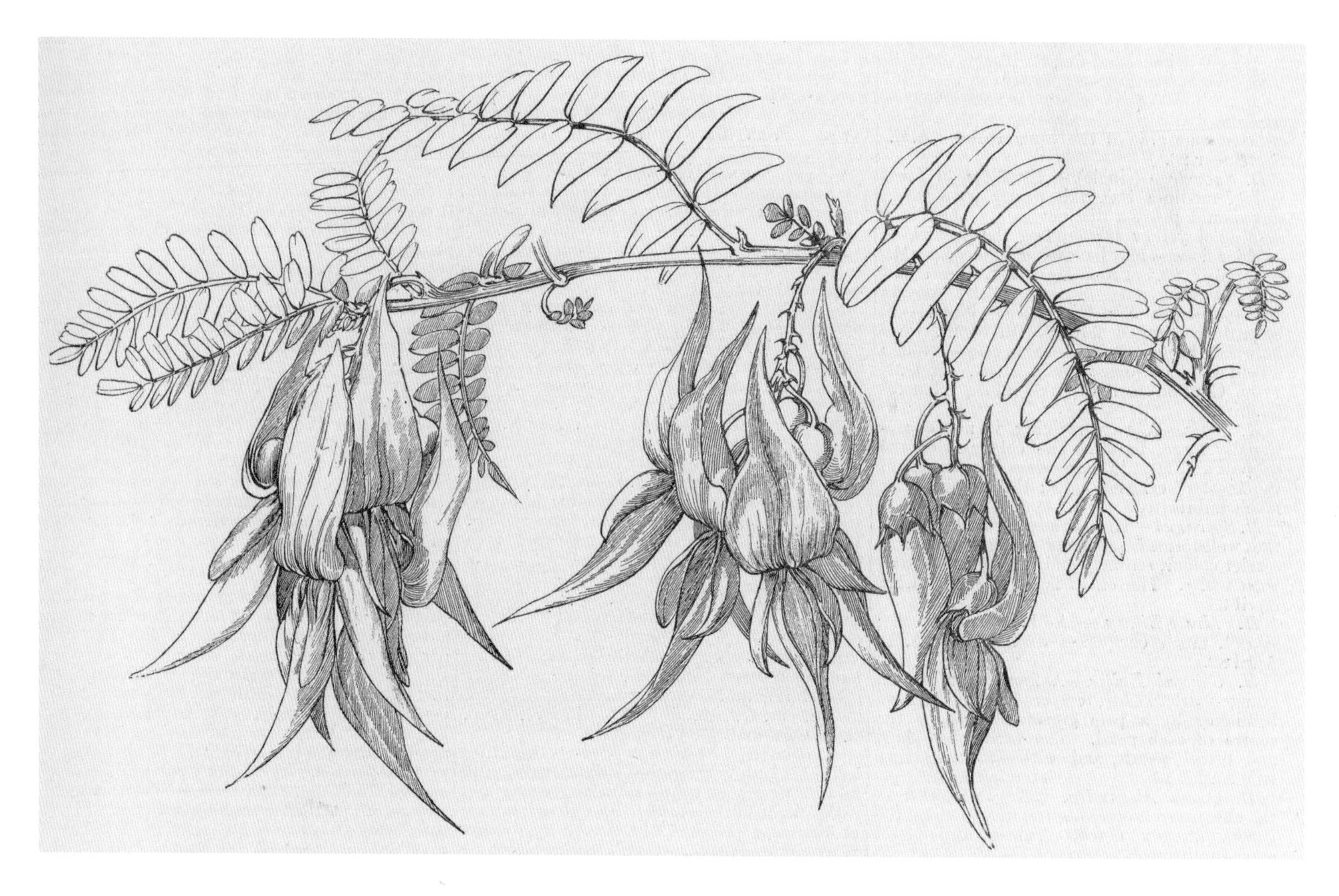

C. puniceus seed pods were also eaten raw and the shrubs cultivated for display.

The rapid decline of *C. puniceus* has sparked a passionate conservation movement, although a coordinated, evidence-based approach will be the only route to long-term success. Research has identified and tagged genetically distinct subpopulations, informing the best sources for raising plants for reintroduction. Having only 150 wild plants means an inherently narrow gene pool, but seeking and perpetuating any genetic diversity marginally increases the resilience of reintroduced plants.

Wild-sourced *Clianthus* are raised in seed orchards, the progeny of known parent specimens. Plants of good provenance have been introduced to roadsides and restored habitats, and the threatening phalanx of exotic grazers is now subject to reduction programmes.

Clianthus seeds, coloured a mottled grey and black and nestled within long pea-like pods, may hold the ultimate answer to the species' survival. Capable of sustained periods of dormancy without losing viability, the seeds have the potential to outlive current threats and regenerate *C. puniceus* populations in the wild.

Regeneration could come through the propagation and planting of banked orthodox seeds, which can survive a drying or freezing process in an ex-situ environment. It could also happen through the wild "seed bank", the repository of grains held within the earth, as the seeds are capable of germinating when conditions are right.

Seeds are tough, resilient systems for ensuring a plant's survival. Utilizing the inherent capacity for dormancy in orthodox seeds through banking buys us time to save, conserve and restore – an insurance policy against extinction.

ABOVE: *Clianthus puniceus* (as *Donia punicea*) from *The Garden*, 1871.

OPPOSITE: *Clianthus puniceus* by S. Watts after Miss Drake, *c.* 1835. Wellcome Collection.

HORT. TRANS. Vol. 1. Pl. 22
SECOND SERIES
Miss Drake. del.
Clianthus puniceus.
S. Watts, sculp.

8

BUSH LILY

Clivia miniata

The great glasshouses of Kew's botanic gardens conjure otherworldly environments. On a cold winter's day, with grey sky and biting wind, the transition into the glasshouse space is transformative: heat, humidity, fragrance, filtered light. The silhouettes of palm and banana leaf, hairy tree ferns, spiky Victoria water lilies.

The skilful horticulturists who conjure this exotic theatre of plants have an uncanny talent for matching a species' wild needs to niches within their new abstract environment. The layers within a complex plant community, from rainforest to Mediterranean woodland, are carefully accommodated into the glasshouse spaces, utilizing height, depth, light and shade. Look carefully under the benches or in places of dappled shade and you'll find a plant both endearingly modest and ravishingly beautiful.

Clivia miniata, the specialized dweller of glasshouse nooks and crannies, is endemic to South Africa and Swaziland, favouring the forest floors of the provinces KwaZulu-Natal, Mpumalanga and the Eastern Cape. A structured array of strap-shaped glossy green leaves arch from an underground storage organ, informally referred to as a bulb, but more accurately a tuberous root. In the right conditions, it flowers profusely, glorious waxy-petalled trumpets in shades of red, orange or yellow, borne on strong fleshy stems. Plant breeders, especially in the Far East, have gradually expanded the *Clivia* palette, bringing in sharp yellows and creams.

Wild bush lily grows in large stands, thriving in the dappled shade of the forest floor. This natural abundance may have encouraged liberal harvesting for the medicinal plant trade, a practice now critically undermining its survival.

OPPOSITE: *Clivia miniata* (as *Imantophyllum miniatum*) from *Album van Eeden, Haarlem's flora, afbeeldingen in kleurendruk van verschillende bol-en knolgewassen*, 1872–81.

ABOVE: *Clivia miniata* from *L'Illustration horticole*, 1893.

The South African National Biodiversity Institute (SANBI) rate *C. miniata* as nationally "vulnerable", an assessment underpinned by an alarming 90 per cent population decline over a long-term survey period.

C. miniata is a highly toxic plant, containing several alkaloid chemicals, including lycorine. A compound also found in other members of the Amaryllidaceae family, such as *Narcissus* (daffodils), lycorine causes nausea, vomiting, diarrhoea and potentially death. Compounds this potent can have valuable medical applications in carefully calibrated doses, and *Clivia* plants have a long history in Zulu medicine.

They have been used to treat snakebite, fever, and urinary complaints, and are applied to accentuate uterine contractions in childbirth. Strongly associated with protection, the plant is used by the Zulu people as a charm against evil. Research on the efficacy and possible clinical applications for these biochemically potent alkaloids is ongoing.

Clivia plants are harvested whole for medicinal use, dug up directly from their forest habitat and traded at market, with little attempt by purveyors to distinguish which species is for sale. An estimated 10,000 kilograms of the plant are sold annually through one Johannesburg market alone. As larger populations become fragmented and lose viability, traders push deeper into the forest, clearing increasingly remote stands.

A shift from heavy exploitation to thoughtful cultivation can support the survival of *C. miniata*. Tuberous plants naturally propagate themselves, regularly producing divisions that can be carefully split from the parent plant without weakening it. By shifting the primary harvesting source of the bush lily from wild forest to cultivated plots, a sustainable path forward for a culturally and medicinally significant plant emerges.

ABOVE: *Clivia miniata* (as *Himantophyllum miniatum*) from *Revue horticole*, 1876.

OPPOSITE: *Clivia miniata and Moths, Natal*, by Marianne North, 1882.

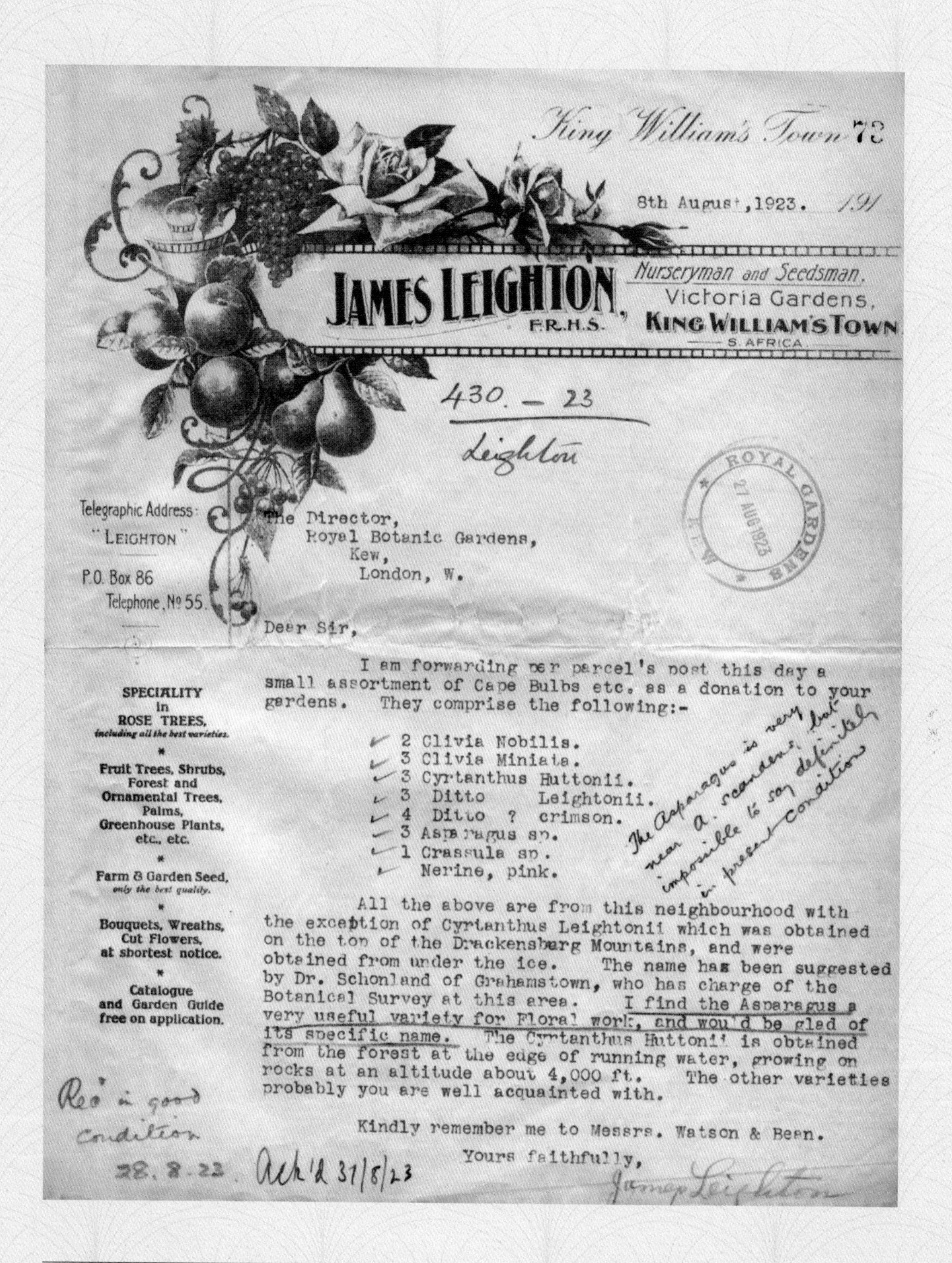

King William's Town 8th August, 1923. 191

JAMES LEIGHTON, F.R.H.S. Nurseryman and Seedsman, Victoria Gardens, KING WILLIAM'S TOWN S. AFRICA

430. – 23
Leighton

ROYAL GARDENS KEW 27 AUG 1923

Telegraphic Address: "LEIGHTON"
P.O. Box 86
Telephone, No 55.

SPECIALITY in ROSE TREES, including all the best varieties.
Fruit Trees, Shrubs, Forest and Ornamental Trees, Palms, Greenhouse Plants, etc., etc.
Farm & Garden Seed, only the best quality.
Bouquets, Wreaths, Cut Flowers, at shortest notice.
Catalogue and Garden Guide free on application.

The Director,
Royal Botanic Gardens,
Kew,
London, W.

Dear Sir,

I am forwarding per parcel's post this day a small assortment of Cape Bulbs etc. as a donation to your gardens. They comprise the following:-

2 Clivia Nobilis.
3 Clivia Miniata.
3 Cyrtanthus Huttonii.
3 Ditto Leightonii.
4 Ditto ? crimson.
3 Asparagus sp.
1 Crassula sp.
Nerine, pink.

The Asparagus is very near A. scandens, but impossible to say definitely in present condition

All the above are from this neighbourhood with the exception of Cyrtanthus Leightonii which was obtained on the top of the Drackensberg Mountains, and were obtained from under the ice. The name has been suggested by Dr. Schonland of Grahamstown, who has charge of the Botanical Survey at this area. I find the Asparagus a very useful variety for Floral work, and would be glad of its specific name. The Cyrtanthus Huttonii is obtained from the forest at the edge of running water, growing on rocks at an altitude about 4,000 ft. The other varieties probably you are well acquainted with.

Kindly remember me to Messrs. Watson & Bean.

Yours faithfully,
James Leighton

Rec'd in good condition 28. 8. 23.
Ack'd 31/8/23

Letter from James Leighton, Victoria Gardens, King William's Town, South Africa, to Sir Arthur William Hill, Director of Kew, 8 August 1923.

Having briefly worked as a gardener at Kew, Leighton was an important figure in King William's Town, where he established a nursery (as per the headed letter above), was curator of the Botanical Gardens, as well as town mayor. Enclosed in this letter he sends various bulbs of *Clivia*, including *C. miniata*, as well as *Asparagus*, *Crassula* and *Nerine*.

OPPOSITE: *Clivia miniata* (as *Imantophyllum miniatum*) by Walter Hood Fitch from *Curtis's Botanical Magazine*, 1854.

HIGHLAND COFFEE

Coffea stenophylla

The genus *Coffea* has provided stimulation globally for millennia. Coffee is widely cultivated with strong associations to many countries: Colombia, Jamaica, Guatemala and Brazil are centres for production, while Italy transformed the West's coffee consumption. Searching for the wild origins of coffee takes us to the highlands of Sierra Leone and Ethiopia.

Coffee is a globally dominant commodity. Valued at over $20 billion (approximately £15 billion) per annum, global estimates for consumption are around 500 billion cups per year. The world's top coffee exporter, Brazil, produced 2.6 million tonnes in 2016, from a swathe of plantations totalling 27,000 square kilometres.

Two plant species are responsible for these extraordinary statistics: *Coffea arabica* (producing refined arabica coffee) and *Coffea canephora* (the more utilitarian robusta). Both are attractive, if understated, evergreen shrubs, with glossy foliage the backdrop to fragrant star-shaped white flowers. Once pollinated, flowers become bright red glossy fruits, the so-called "cherry" that's washed and dried to make coffee beans.

BELOW: Flowers and beans of *Coffea arabica* by Manu Lall, an example of Company Art commissioned from Indian artists by British East India Company botanists, nineteenth century.

OPPOSITE: *Coffea stenophylla* by Matilda Smith from *Curtis's Botanical Magazine*, 1896.

LEFT: *Coffea arabica* from Elizabeth Blackwell, *Herbarium Blackwellianum*, 1760.

When coffee production is threatened, the economic power of this crop causes governments to take notice. *C. arabica* is becoming increasingly difficult to grow in its heartlands due to a combination of fungal rust, the coffee borer beetle and climate change.

A combination of biotic (living) and abiotic (environmental) stresses holds particular threats for *C. arabica*. The delicate taste of its beans are matched by a delicate physiology, a stark contrast to the robusta (*C. canephora*), whose named is well-earned. *C. arabica* is a wild hybrid less than 50,000 years old and the cultivated plants derived from it have little genetic variability. Major perturbations to this plant resonate across its global population.

C. arabica originates from the south-western highlands of Ethiopia. This country is home to a multitude of subtly different cultivated forms, many varying little from the wild type. Climate change is a stark reality for Ethiopian coffee farmers, with average annual increases of 1.3°C between 1960 and 2006 having alarming effects on their crops. *C. arabica* is showing significant signs of drought stress in the lowest altitude plantations, reducing productivity. The Royal Botanic Gardens, Kew has worked closely with local coffee growers, anticipating future climate niches to suggest higher-altitude plantations that may be better insulated from future environmental stress.

Reliance on *C. arabica*, a delicate plant under a range of threats, to continue producing the world's coffee could be

DEPARTEMENT VAN LANDBOUW, NIJVERHEID EN HANDEL. BUITENZORG (JAVA).
SELECTIE-STATION
№ 349

Bangelan, 12 th June, 1915.
BUITENZORG, 293 191 .
№

Mr. Arthur W. Hill,
Assistent-Director of the
Royal Gardens Kew.

ROYAL GARDENS KEW

Dear Mr. Hill,

I am in receipt of your letter dated 4 V which reaches me here in the experimental garden for coffee, placed recently under my supervision.

I have little opportunity here to enter fully upon the subject of a systematical study of Coffea. I think our work here - selection, working with living plants one generation after the other - should keep in close touch with the systematical work in scientific centra abroad. I will propose a scheme for it as soon as I am back in Buitenzorg and hope that the Director of Kew will approve it.

Noted. [illegible]

Of course I will be glad to send herbarium specimens to Kew. But do not trust the names of botanical gardens, especially in the tropics; and do not forget that many of their varieties are hybrids. I feel sure f.i. that De Wildeman made species out of coffea's, which are botan. garden-hybrids. I think his Coffea affinis should be Coffea stenophylla x liberica hort.

Therefore I want to divide my species of Coffea into larger sections, calling each section by a name familiar to the planter, as arabica, liberica, robusta, etc., and giving then numbers to the different varieties, smaller species and races among them. Under liberica f.i. I would put: abeokutae, Dewevrei, aruwimiensis (this name does no chairman of a planters union pronounce).

Of each number I would send a complete sample to the

DEPARTEMENT VAN LANDBOUW, NIJVERHEID EN HANDEL. BUITENZORG (JAVA).
SELECTIE-STATION
№

BUITENZORG, 294 191 .
№

pal herbaria, Kew in the first line. The systematist can make out then, what the correct name is; for our - mostly practical - work this is of little importance.

To send you regularly those samples is what I hope to do for Kew. What Kew can do for me is this. I start my selection as much as possible with seeds from wild plants, collected for each tree apart so that my races all descend from one original wild tree. This will help to avoid the confusion hybridisation causes in the group. From the Belgian Congo and from french colonies I received coffeeseeds collected in that way, from English colonies up till now none. I would be glad if Kew could help me to get also from your african protectorates seeds from wild coffee trees, collected for each seed bearer separately.

I hopeto start a special garden next year for this work. I am not sure - it would cost the government about £ 2000 per annum, and war has a bad influence on peaceful work, even with neutrals.

As soon as I am back in Buitenzorg - in a month or so - I hope to write you more fully on the subject.

Trinidad left pleasant souvenirs with me. I often hear from Rorer, we exchange seeds and plants. Will you not come out to the East? I think things like our garden here would interest you, and I would be very glad to show it to you.

With kind regards,

yours very truly, P. J. S. Cramer

Chief Plant breeding station.

Letter from Pieter Johannes Samuel Cramer, Chief Plant Breeding Station, Bangelan, Indonesia, to Sir Arthur William Hill, Assistant Director of Kew, 12 June 1915.

Cramer writes back to Hill from the experimental garden for coffee in Bangelan, laying out his proposal for sending samples to Kew and his views on the current coffee nomenclature coming out of botanic gardens.

Pl. XXIII.
2.
3.
1.
4.
5

OPPOSITE: *Coffea arabica* by H. Sowerby from Edward Hamilton, *Flora Homoeopathica*, 1852.

LEFT: *Foliage, flowers, and fruit of the Coffee, Jamaica*, by Marianne North, 1872.

unsustainable. Could we build greater economic resilience by searching the diversity of the 120 species-strong *Coffea* genus for future crops? Across Africa and Asia, other species of *Coffea* are made into coffee, often in highly localized and specific systems, including an as-yet-unnamed species unique to Mozambique.

Highland coffee is indigenous to the West African countries of Guinea, Sierra Leone and Côte d'Ivoire. Coffee brewed from this elusive plant is delicious, regarded by some as superior to *C. arabica*. In the search for a resilient coffee-growing economy, *Coffea stenophylla*, either in cultivated form or interbred with other species, can be part of a more diverse solution. Or it could be, if this elusive plant was easier to find.

In 2018, Kew scientist Dr Aaron Davis set out to achieve what previous expeditions had failed to do: find wild *C. stenophylla* in Sierra Leone. Forest clearance for timber or agriculture is rapidly fragmenting this species across its wider distribution, earning an IUCN Red List rating of "Vulnerable". Davis eventually found the elusive wild stands across several locations, but they were isolated and fragmented.

Conserving the world's biodiversity can provide us with economic wealth and resilience. Species capable of providing future crops, fuels, medicines and fibres can benefit us only if we don't unwittingly drive them to extinction. A world with no coffee is unimaginable for some, a spur to conservation action as stimulating as the crop itself.

10

DRAGON TREE

Dracaena draco

As a relic of an ancient flora, a "tree" actually more closely related to asparagus (in a form so striking that even the most plant-ambivalent tourist scrambles to capture it on their smartphone), and a star of Greek mythology, the dragon tree is a compelling ambassador for the plant world.

Even with its celebrity status and vivid role within Macaronesian island folklore, the dragon tree is not immune to threats and is now rated as "Vulnerable" by IUCN. With declining indigenous distributions, the dragon tree is widely cultivated, and is a staple specimen of Mediterranean-climate botanic gardens around the world.

The dragon tree is a monocotyledon: a botanical classification deriving from the production of a single seedling leaf but linked to a wide range of structural traits (parallel leaf veins, flower parts usually in threes, fibrous roots and scattered vascular bundles). The adult leaves are stiff and spiky, with a glaucous tinge.

Despite its appearance, it's not a true tree. The thick, gnarled trunk is partly composed of aerial roots, which appear to play a role in supporting the vast girth (up to 8 metres) developed in mature specimens. The trunk grows for the first 10–15 years of the tree's life, and then terminates to form the plant's first flower. From this flower comes a branch, from this branch come further branches and thus, this convoluted pattern of growth creates the dragon tree's extraordinary umbrella-like structure.

Dracaena draco reflects its origins – in flora living millions of years ago – with the distribution of the species on both the African mainland and the Macaronesian islands off Africa's Atlantic coast. Distinct populations are recorded in Madeira and five of the seven Canary Islands, and these were recently complemented by the discovery of a new subpopulation in the Anti-Atlas mountains of Morocco. DNA from disparate populations of a species can be analyzed to create a phylogenetic tree, highlighting points of divergence and convergence from a common ancestor.

The dragon tree's vivid red resin, known evocatively as "dragon's blood", has been used in mummification, violin staining and to prevent tools rusting. The resin had a period as a commercially traded commodity, with 76 cases of

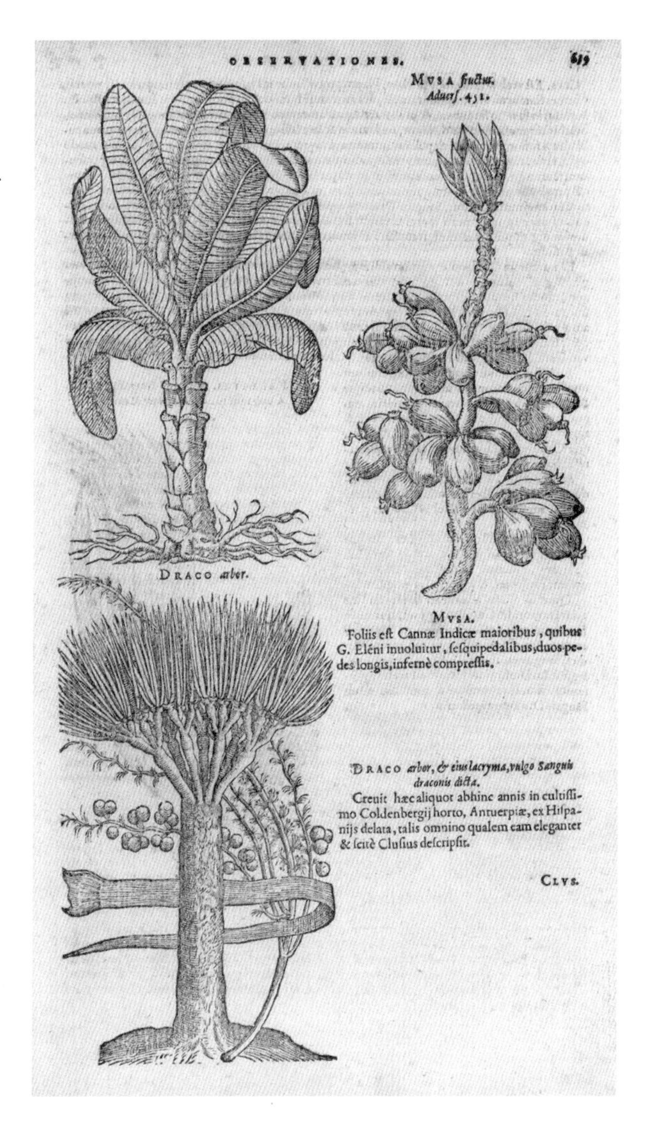

OBSERVATIONES. 619

MVSA *fructus.*
Aduerſ. 451.

DRACO *arbor.*

MVSA.
Foliis eſt Cannæ Indicæ maioribus, quibus G. Eléni inuoluitur, ſeſquipedalibus, duos pedes longis, infernè compreſſis.

DRACO *arbor, & eius lacryma, vulgo Sanguis draconis dicta.*
Creuit hæc aliquot abhinc annis in cultiſſimo Coldenbergij horto, Antuerpiæ, ex Hiſpanijs delata, talis omnino qualem eam eleganter & ſcitè Cluſius deſcripſit.

CLVS.

OPPOSITE: *Dragon Tree at Orotava, Teneriffe*, by Marianne North, 1875.

ABOVE: *Dracaena draco* from Carolus Clusius, *Rariorum plantarum historia*, 1601.

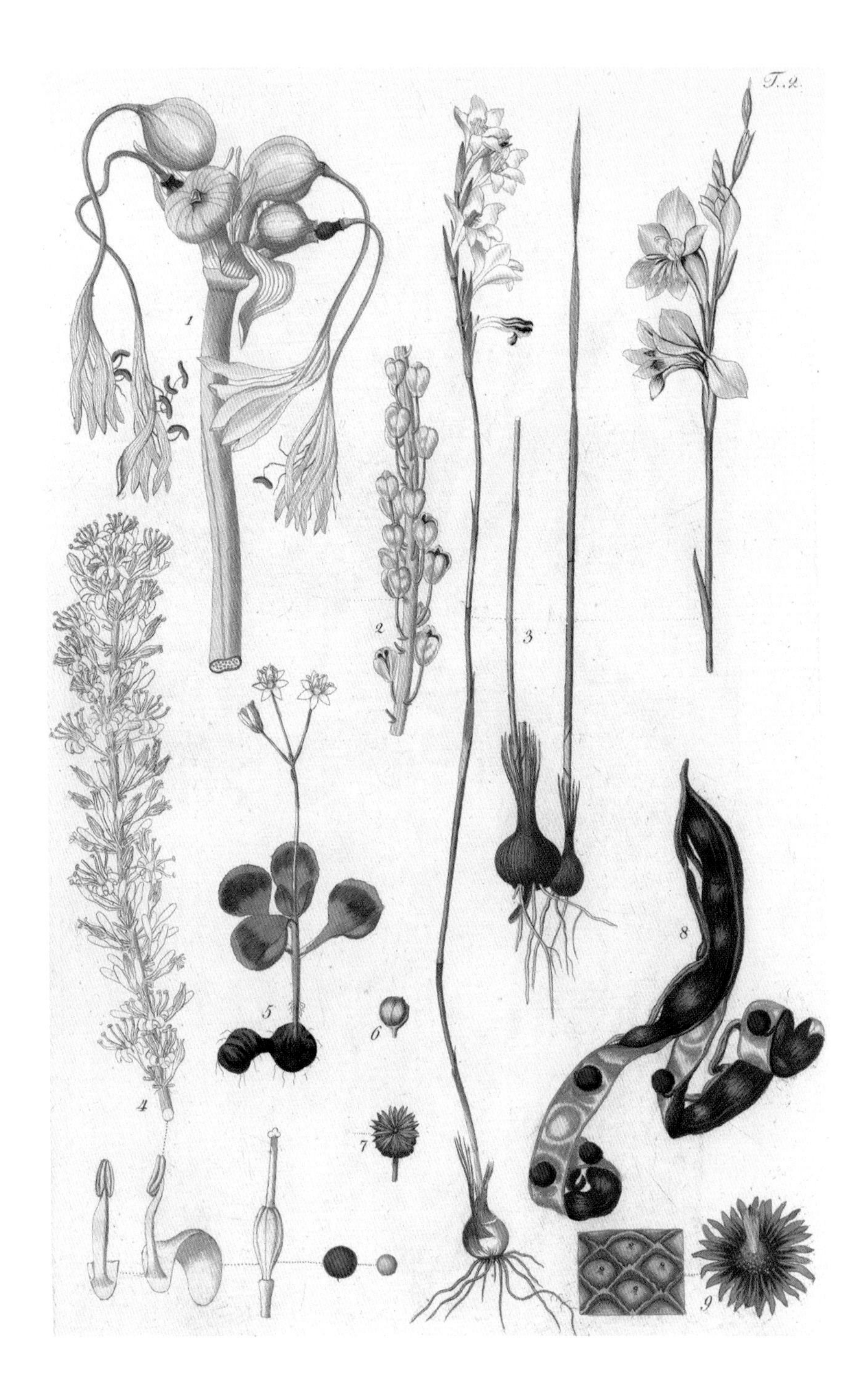

LEFT: *Dracaena draco* (no. 4) from Nikolaus Joseph von Jacquin, *Fragmenta botanica, figuris coloratis illustrata*, 1809.

dragon's blood shipped by Brown Brothers and Company famously heading to the bottom of the Atlantic in the hold of the *Titanic*.

Greek myth links the arrival of the dragon tree to the eleventh labour of hard-working legendary hero Hercules. As he slew the fearsome multi-headed dragon Ladon, wherever the ensuing flow of blood settled, a multi-headed dragon tree would rise. It's not hard to imagine how the imposing, unearthly mature dragon trees inspired myth-making.

Despite its status as a national symbol of the Canary Islands, dragon trees have been cleared for agriculture, especially cattle ranching. Young dragon trees are palatable to domestic goats, and unprotected plants are destroyed by uncontrolled grazing. While tolerant of periods of drought, the dragon tree has evolved to draw moisture from low-lying clouds and mists. Climate change has reduced the occurrence and endurance of these moisture sources, increasing environmental stress.

These external pressures are conspiring against a plant with inherently low viability. The dragon tree was dependent on a mutually beneficial partnership with a flightless bird endemic to the Canary Islands, which distributed seeds after eating the plant's hard fruits. With its avian partner long extinct, the dragon tree is now reliant on the islands' human population to colonize new niches – a fragile dependence indeed.

Letter & thanks sent.

Sitio del Pardo
Puerto de Orotava, Tenerife
20th May 1875

My dear Sir,

In the hope that it may not be wholly without botanical interest to you, I have ventured to send you a section of a Dragon tree (Dracaena Draco) which fell suddenly about a week ago in my garden.

I shipped it three days since by the Steamer "Risca" from this Port, and it should reach you in about a fortnight or three weeks.

You will, I hope, soon see a sketch of the tree made by that accomplished lady Miss North, who has just been on a visit to Tenerife & whose beautiful sketches are, I know, appreciated by you.

The age of the tree I think may be from ninety to a hundred years, as my old gardener, himself just the age of the century, remembers it when he was a little boy to have been some six or eight feet high.

Its dimensions at present were about 9 feet in circumference at the ground & 3 feet at the top of the trunk where the branches spread out.

Its height from the ground to the spreading out of the branches was about 18 feet, and thence to the top of the branches some 9 or 10 feet more.

Apologizing for the freedom I have taken though personally unknown to you, I remain,

My dear Sir
Very truly yours
Chas. Smith

To Dr. Hooker
Kew Gardens.

Letter from Charles Smith, Sitio del Pardo, Puerto de Orotava, Tenerife to Sir Joseph Dalton Hooker, Director of Kew, 20 May 1875.

"In the hope that it may not be wholly without botanical interest to you..." Charles Smith begins his letter to Hooker, in which he has included a section of a *Dracaena draco* tree which used to be in his garden. He believes the tree to have been 90 to 100 years old, particularly as his 100-year-old gardener remembers it as a boy. Before it was destroyed in a storm, Marianne North painted the tree, depicted in her work *Dragon Tree in the Garden of Mr Smith, Teneriffe.*

ABOVE: *Dracaena draco* from Elizabeth Blackwell, *Herbarium Blackwellianum*, 1760.

OPPOSITE: *Dracaena draco* from Johann Wilhelm Weinmann, *Phytanthoza iconographia*, 1745.

N.785.
b
a
b
a. Palma prunifera fo=
liis Yuccæ, fructu ra=
cemoso sanguinem dra=
conis fundens, Arbre du
Dragon, Drachen=
blutbaum.
b. Palma Guineensis vinifera.
H.

11

SMOOTH PURPLE CONEFLOWER

Echinacea laevigata

The smooth coneflower – a relative to the source of a globally popular cold remedy, the supplier of a rich gene pool to horticultural plant breeders, and a symbolic cornerstone of wild American grassland – is under threat in its homeland. *Echinacea laevigata* is distributed across the south-eastern United States, in a wild plant community called the Piedmont Prairies. Found in just four states – North Carolina, South Carolina, Georgia and Virginia – the smooth purple coneflower has been reduced to small fragmented populations. Consequently, it is listed as federally endangered, requiring protection by state legislation and support from regional conservation programmes.

The smooth coneflower is a beautiful, elegant component of the Piedmont Prairies. Borne on strong wiry stems, the flower has slender pale pink ray florets (the petals that radiate from the centre of the flower, like the rays of the sun) and dense, dark purple disk florets. It grows through clump-forming grasses in woodland glades, using flower stems up to 1.5 metres tall to reach available light. The open, accessible nectaries in the flower attract a wide range of pollinating insects, from bees to bugs.

This flower's fortunes have been firmly aligned with human impact. Prairies are complex assemblages of non-woody forbs (flowering plants) and grasses, often fringed by scrubby trees such as the loblolly pine (*Pinus taeda*). These rich compositions, often filled with over a hundred species, are not inherently stable and can revert to woodland through the process of succession. In this situation, stress and disturbance are valuable forces, stopping the slow creep of more aggressive plants, and ultimately that of tree seedlings as well. The two most potent sources of "good" stress are fire and grazing.

RIGHT: *Echinacea purpurea* (as *Rudbeckia purpurea*) from *Curtis's Botanical Magazine*, 1787.

OPPOSITE: *Echinacea purpurea* (as *Echinacea intermedia*) from Louis van Houtte, *Flore des Serres et des Jardins de l'Europe*, 1848.

ABOVE: *Echinacea purpurea* (as *Rudbeckia purpurea*) from Sydenham Teast Edwards, *The new botanic garden*, 1812. A yellow rose is featured on the right of the illustration.

OPPOSITE: *Echinacea laevigata* from Mark Catesby, *The Natural History of Carolina, Florida, and the Bahama Islands*, 1754.

Fire has been managed to shape large-scale landscapes such as prairies for millennia. Prairie plant species have the ability to re-sprout after burning, making fire a potent tool to regenerate vegetation. The young foliage attracted roaming bison, a desirable prey for the Native Americans who stewarded these lands, while offering fresh sources of food, medicine and fibres. The grazing of large bison herds roaming freely across the prairie placed the more vigorous plant species under pressure, levelling the playing field and driving diversity.

E. laevigata is a plant of open woodland glades, and it needs bright, direct sunshine and damp around its roots. With changing land use comes the loss of good stress and disturbance that maintained the delicate status quo of species diversity. Without carefully stewarded fires, tree seedlings filled the smooth coneflower's desired glades. Without palatable young foliage, roaming grazers no longer maintained the balance between species. Fragile due to its precise niche requirements, this flower is further weakened by over-collection, herbicide use, and competition from invasive species.

Conservation programmes endeavour to stabilize the decline of the smooth coneflower. Botanic gardens such as the Sarah P. Duke Gardens in North Carolina have created stunning horticultural displays using plants native to their state. Gardens like these are capable of inspiring local residents to support their state flora while offering a valuable ex-situ source of seed for conservation. The sight of the slender, beautiful smooth coneflower growing en masse poses a simple question to visitors: "How can we let this plant become extinct?"

This species reminds us that plant conservation needs to happen everywhere, not just far away in burning rainforests. The pressures threatening biodiversity – climate change, shifting land use and new pests and diseases – endangers species from Oxfordshire to Virginia, from Brazil to Malaysia. The principles of conservation, which include identifying species, quantifying the extent and health of existing populations, evaluating threats and monitoring long-term viability, apply everywhere.

Dr Hooker
Professor of Botany
Glasgow

16/480(21
32
160
16
8/400/500
7

Town of Velasco Mouth of the Rio Brazos Texas 90
May 14th 1833
168
Dear Sir

Letter from Thomas Drummond, Town of Velasco, Texas, USA, to Sir William Jackson Hooker, Director of Kew, 14 May 1833.

Drummond was a Scottish botanist who endured numerous hardships mapping and collecting flora and fauna in North America. Here he describes to Hooker the illness and delay that have impacted his work, he himself falling ill with, and recovering from, cholera where several others died. Despite the difficulties endured, he has collected 100 plant species, 60 bird species, two snake species and several land snails. His next trip will take him across the Colorado River. Drummond died in Havana, two years after writing this letter, possibly from septicaemia.

OPPOSITE:
Echinacea angustifolia by Walter Hood Fitch from *Curtis's Botanical Magazine*, 1861.

1.
2.
3.
W. Fitch, del. et lith.
Vincent Brooks, Imp.

ENSET

Ensete ventricosum

The family Musaceae, more commonly known as banana, is composed of three genera: *Musa*, *Musella* and *Ensete*. Two fruits hailing from this family dominate global agriculture: sweet banana is the most traded fruit in the world, while 45 million tonnes of starchy, savoury plantain are produced annually. Bananas bring lush, exotic qualities to horticultural display, with their towering pillars of copper or emerald with vast leathery leaves.

Despite this towering stature, bananas are not trees but herbs with no true woody tissue. Growing between 6 and 20 metres (the towering *Musa ingens* claims the title of world's tallest banana plant), bananas are early jungle or woodland colonizers. Growing rapidly to exploit new-found light in cleared vegetation, many banana species live fast and die young, producing large volumes of seed as their final act.

Musa basjoo is a banana capable of expanding the British gardener's horticultural horizons. Frost-hardy up to -15°C, it will comfortably survive winters in all but the coldest gardens, shrugging off any damage from the colder months with strong new spring foliage. When *Musa basjoo* is combined with other tough-but-exotic species such as *Tetrapanax papyrifer* (rice paper plant) or *Hedychium gardnerianum* (kahili ginger), the exotic sub-tropical garden style can be explored with minimal risk, and it works particularly well in urban settings.

The enset or false banana (*Ensete ventricosum*) is an abundant component of sub-Saharan forests, classified as of "Least Concern" in IUCN's Red List because of its stable population. As an agricultural crop, enset is surprisingly rare in cultivation. A richly nutritious plant, it's only grown for food in the sophisticated agro-forestry systems of the highlands of southern Ethiopia, where perennial and annual plants are skilfully combined to produce year-round food.

Enset is not a plant to cultivate for its fruit. Peeling the skin back reveals no sweet flesh, just hard black seeds. But this is a banana with other desirable traits. Enset is a source of plentiful starchy calories, which are exploited and extracted through ingenious cultivation. As a hapaxanthic plant, enset flowers and sets seed once before dying, and harvesting must take place before this energy-sapping event.

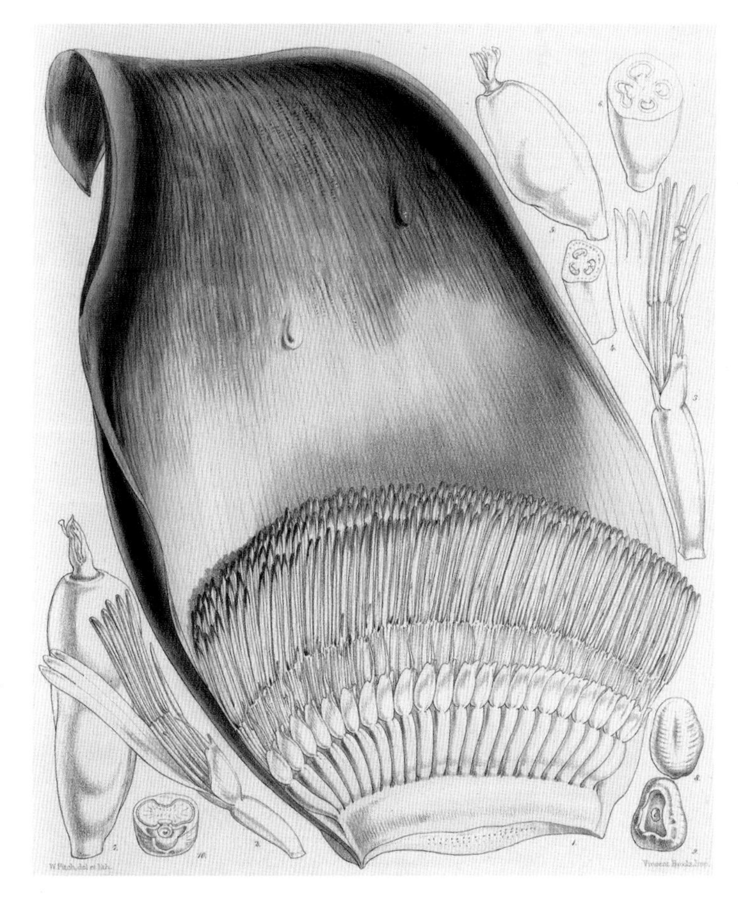

OPPOSITE: *Ensete ventricosum* (as *Musa ensete*) by Walter Hood Fitch from *Curtis's Botanical Magazine*, 1861.

ABOVE: *Ensete superbum* (as *Musa superba*) by Walter Hood Fitch from *Curtis's Botanical Magazine*, 1861.

Leaves are stripped and the plant is carefully prised from the ground, the beachball-sized underground storage organ known as a corm carefully separated.

The multi-layered stem is gently deconstructed, each layer containing a calorific deposit of starch. Stems are scraped clean of starch and the corm is mashed. Both are then fermented in pits rich in microbiology that cure the starch into "kocho", a staple food. This is just one of a range of valuable enset products, which also includes fibres, cattle fodder, packaging, and the versatile whey-like liquid called "bulla". Enset is propagated by transplanting strong suckering growth from parent plants, a system that clones desirable traits from one generation to the next.

LEFT: *Ensete ventricosum* (as *Musa ensete*) from Louis van Houtte, *Flore des serres et des jardin de l'Europe*, 1859–61.

BELOW: *Abyssinian Ensete in a garden in Teneriffe*, by Marianne North, 1875.

Letter from Walter Fox, Botanic Gardens, Singapore, to Sir William Thiselton-Dyer, 12 December 1894.

Fox is very pleased to have received a box of *Ensete ventricosum* (referred to here as *Musa ensete*) seeds, due to the "fine foliage" of this species which will far-outrival that of *Musa superba* (now *Ensete superbum*), of which they have many. In return Fox is sending to Kew a box of *Musa sumatrana* (now *Musa acuminata* var. *sumatrana*) seeds, known locally as "pisang karok".

Fox worked at the Botanic Gardens, Singapore for over 30 years, after which he ran the Waterfall Gardens in Penang, Malaysia, until his retirement in 1910.

OPPOSITE: *Ensete ventricosum* (as *Musa ensete*) from Eduard Regel, *Gartenflora*, 1870. *Loeselia mexicana* (as *Loeselia coccinea*) is featured on the left.

19
1895.

ackd. 16.1.95.

Botanic Gardens
Singapore
12th December 1894

Dear Sir,

I have to thank you for the box of very fine plants of the various Phyllocacti which arrived in very fair order. Some of the leaves had rotted but I dont think I shall lose one.

The box of "Musa Ensete" seeds arrived yesterday. I am very pleased to get such a very fine foliage plant as this is. We have of course lots

Musa superba & effective as this is "M. Ensete" will far outrival it.

By this mail I am sending you a box of "Musa Sumatrana" seeds – the Pisang Karok of the Malays. I am also getting up for pots some young plants of "Ficus alba" as requested by Dr. Morris; it is a curious fact that many of our commonest plants are impatient of pot culture.

I have just heard unofficially that the somewhat sweeping reforms in our Dept. proposed by the Retrenchment Commn. are to say the least to be somewhat modified. Mr. Ridley will I suppose return if not to his old appt. to some modification of it — & Mr. Derry go back to Malacca but as the Official papers have not yet reached me I do not know anything very definite.

I am dear Sir
Yrs faithfully
Walter Fox

W. T. Thiselton-Dyer Esq.

Recd 18.1.95

1.
2.
3.

13

THE GENUS *EUCALYPTUS*

With its shimmering silver foliage, shaggy or plated bark and an instantly recognizable menthol aroma, few trees are as redolent of the southern hemisphere as those of the genus *Eucalyptus*. While its origins are Antipodean, *Eucalyptus* now has a global impact (it's sometimes referred to as "the world's most planted tree"), dividing opinion as it invades foreign lands while some species have become endangered on home soil.

The genus *Eucalyptus* is huge and contains over 800 species. Spanning the Australian mainland, Tasmania, New Guinea, the Philippines and Java, it's a wonderfully diverse group, taking forms from towering forest trees to scrubby shrubs. The genus is particularly resistant to fire, with a number of strategies for post-burn regeneration, an integral element of Australian plant ecology. It's a source of fuel, food, medicine and fibre, boasts the world's tallest flowering tree and served as the "canvas" for early Aboriginal painting.

The concept of an endangered *Eucalyptus* will seem alien to conservationists on the African continent and countries with Mediterranean climates seeking to eliminate choking swathes of the tree. *Eucalyptus grandis* (flooded gum) and *Eucalyptus camaldulensis* (river red gum) show highly invasive behaviour in South Africa, threatening the biodiverse Fynbos flora of the Western Cape. Introduced for its rapid growth and wide range of economic uses, *Eucalyptus* sharply splits opinion.

Across East Africa, *Eucalyptus* has been used for telegraph poles, plywood, fuel wood and scaffolding, while its mentholated essential oils are used in perfumery and cold cures. The willingness of *Eucalyptus* to grow rapidly and regenerate after cutting imbues it with great economic value, but its ability to colonize disturbed ground more rapidly than native trees makes it the bane of conservationists. In the fragile, highly biodiverse biome of Madagascar, large-scale seeding and planting of *Eucalyptus* is a serious threat to the island's endemic flora.

But not all *Eucalyptus* trees are thriving – *Eucalyptus morrisbyi* (Morrisby's gum) is in trouble. A diminutive tree that reaches 12 metres at its tallest, it is now only found in two locations in its native Tasmania, earning it an IUCN

ABOVE: *Blue Gum Trees, Silver Wattle, and Sassafras on the Huon Road, Tasmania*, by Marianne North, 1880.

OPPOSITE: *Eucalyptus globulus* from Köhler, *Medizinal-Pflanzen*, 1887.

1
2
3
4
5
6
7
8
9
10

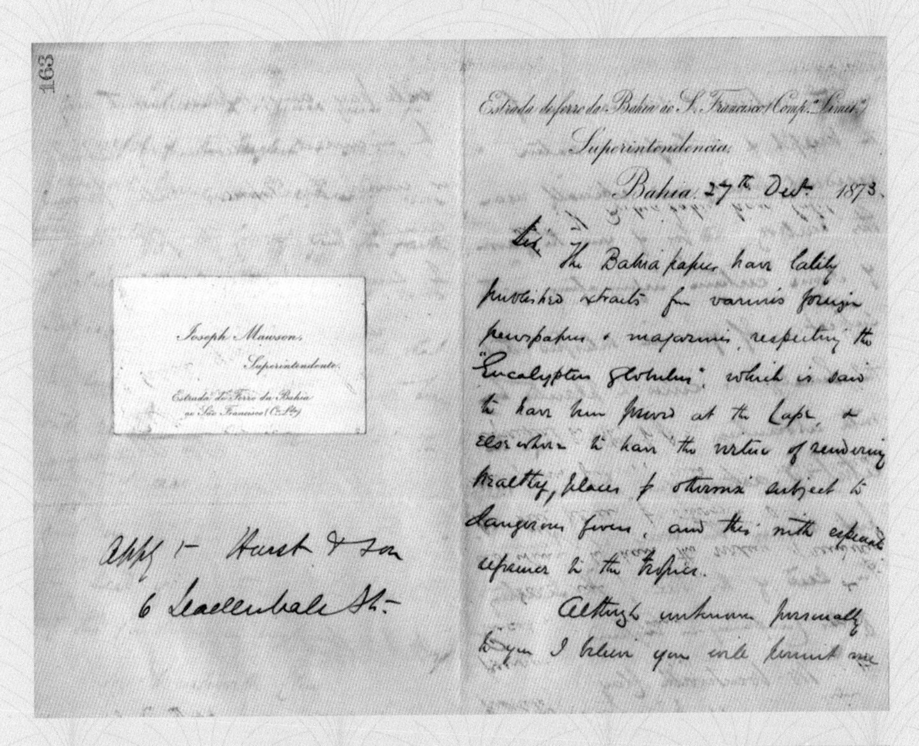

163

Estrada de ferro da Bahia ao S. Francisco Comp.y Limited

Superintendencia

Bahia 27th Oct. 1873.

Sir

The Bahia papers have lately published extracts from various foreign newspapers & magazines respecting the "Eucalyptus globulus" which is said to have been proved at the Cape & elsewhere to have the virtue of rendering healthy places of otherwise subject to dangerous fevers, and this with especial reference to the tropics.

Although unknown personally to you I believe you will permit me

Joseph Mawson,
Superintendente.
Estrada de Ferro da Bahia ao São Francisco (C. L.d)

Apply to Hurst & Son
6 Leadenhall St.

in the interest of science and for the benefit of the English & native residents here, and especially near this railway, to beg of you the favour of some certain information on the subject. If you are satisfied that the Eucalyptus cannot be planted here with advantage I will take the liberty of asking you to send, or put me in the way of obtaining the seed of the tree. The Secretary of this Company in England Mr. Wentworth Clay

will pay any expenses which may be incurred [illegible] or under [illegible] will forward any thing you may be good enough to send [illegible]

I remain Sir
Your very obedt servt
Jos Mawson

Dr J. D. Hooker F.R.S.

Letter and business card from Joseph Mawson, Superintendent, Estrada de ferro da Bahia ao S. Francisco Company Ltd., Bahia, Brazil, to Sir Joseph Dalton Hooker, Director of Kew, 27 October 1873.

In his letter to Hooker, Mawson notes that the Bahia newspapers are reporting on the medicinal properties of *Eucalyptus globulus*, discovered at the Cape and elsewhere. He requests that some *Eucalyptus* be sent, if Hooker believes that it could be successfully planted in Bahia.

OPPOSITE: *Eucalyptus camaldulensis* (as *Eucalyptus rostrata*) from John Ednie Brown, *The Forest Flora of South Australia*, 1892–93.

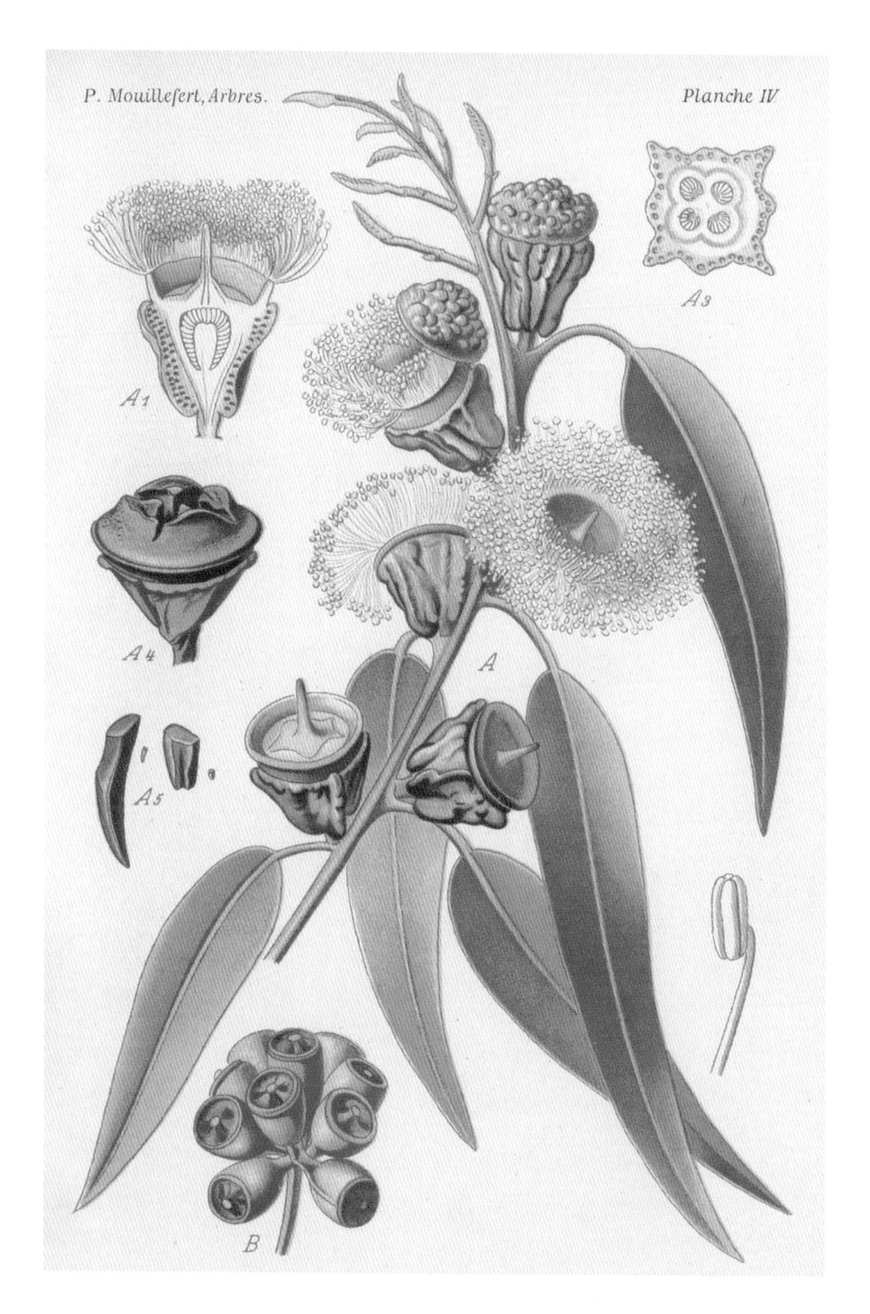

rating of "Endangered". These populations are far from robust: 11.5 hectares of the Calverts Hill Nature Reserve holds just under 2,000 trees, while the other locations are severely fragmented, comprising less than a hectare in total and holding no more than 200 Morrisby's gum trees. Just over 2,000 individual specimens, these are the sole buffer between *E. morrisbyi* and extinction in the wild.

Climate change is slowly eroding this tree's resilience. Its favoured habitat is on the damp side, and as hotter, drier summers make terrain more hostile, *E. morrisbyi* is retreating to wetter gullies. On the margins of its distribution, further pressure from grazing, insect predation and land-clearing for agriculture is taking this species to the edge, making the protection of the Calverts Hill Nature Reserve and global conservation programmes essential. Thanks to coordinated seed collection and propagation, species of *Eucalyptus* are now growing in botanic gardens such as Wakehurst, strengthening the potential for an effective reintroduction.

The story of *Eucalyptus* highlights the power of environmental conditions to change the function of plants, and our responsibility to plant trees with an evidence-based understanding of ecology. Away from biotic (pests, diseases) and abiotic (climate, geology, water availability) constraints, *Eucalyptus* becomes pioneering, rampant and invasive. Woven into its native environment, it is constrained and fragile, threatened by climate change and grazing. The responsibility of ecological balance and stopping species loss sits solely with us.

ABOVE: *Eucalyptus globulus* (figure A, along with *Eucalyptus robusta*, figure B) from Pierre Mouillefert, *Traité des arbres et arbrisseaux*, 1892–98.

OPPOSITE: *Foliage and Flowers of the Blue Gum, and Diamond Birds, Tasmania*, by Marianne North, 1880.

14

COMMON ASH

Fraxinus excelsior

The ash is a cornerstone species in European woodlands, the pioneering opportunist that colonizes disturbed ground with thousands of seedlings and initiates woodland succession, slowly ceding ground to longer-lived trees like beech and oak. As woodlands mature, a handful of ash trees remain: veteran mother trees ready to offer a fresh crop of seedlings should the canopy open again.

When reports of a new disease called ash dieback (caused by the fungus *Hymenoscyphus fraxineus*) emerged in 2013, there was a chilling sense of inevitability about its arrival and impact. Spreading via windblown spores from Asia, it devastated wild ash stands across Europe, travelling across the channel and infecting British woodlands from 2014.

Young ash trees have smooth bark and so readily host lichens, which decorate the tree's trunk with their pale glaucous discs. As the tree matures, the bark slowly fissures, eventually forming a rugged, rough outer covering with superficial similarities to oak. The form of a mature tree is assuredly beautiful: balanced and compact, with swooping low branches. Students challenged to identify "winter twigs", an arduous rite of passage for horticultural trainees, take solace in the ash's immediately identifiable coal-black buds.

Ash has compound leaves: leaflets, or pinnae, borne in opposite pairs along a form of stem called the rachis. This botanical feature allows ash to contribute a glorious marbled light through woodland, one of the true wonders of an early summer country walk.

Prior to the onset of ash dieback, the ash was abundantly distributed across Europe and beyond, from Great Britain to Turkey. Ash's adaptability to a broad range of ecological conditions means it's often the first tree to arrive in disturbed ground and you'll see ash trees growing everywhere from river banks to upland slopes. The tree displays less vitality in climatic extremes, struggling in very hot dry summers and cold winters.

Ash is a remarkable wood: strong, flexible and intensely calorific (releasing a large amount of heat energy when burned). It confers strength and durability to tool handles, and to car frames such as those of Morgan sports cars, while furniture makers accentuate its fine grain and pale wood. The

OPPOSITE: *Fraxinus excelsior* from Otto Wilhelm Thomé, *Flora von Deutschland*, 1886–89.

ABOVE: *Fraxinus excelsior* by Auguste Faguet from Louis Figuier, *The Vegetable World*, 1867.

1
2
3
6
4
5

OPPOSITE: *Fraxinus excelsior* from François Pierre, *Flore Médicale Décrite*, 1815–20.

RIGHT: Herbarium specimen of *Fraxinus excelsior* collected by R. F. Hohenacker in 1838, held at the Royal Botanic Gardens, Kew.

name "ash" is thought to derive from the Anglo-Saxon word "æsc" – meaning both "ash tree" and "spear", as the wood was used to make weapon shafts – and Norse mythology depicts Thor hunting with a trusty ash-handled weapon.

In rural meteorology, the tree has a key role: there's no need to consult long-term weather forecasts if your ash tree comes into leaf before its neighbouring oak. Rustic wisdom indicates we're in "for a soak" (wet summer) in this scenario, or "in for a splash" (dry summer) if foliation is reversed and the oak leaves up first.

The speed and devastation of ash dieback is reflected in ash's IUCN rating. In just a few years, one of our most abundant trees has become "Near Threatened", an unimaginable scenario for a tree regarded by some as "weedy", and the most willing colonizer of bare ground. Trees infected with ash dieback quickly lose the species' characteristic vigour. Those broad, abundant crowns recede, foliage thins and pales, and dead branch tips become a dominant characteristic. Degrees of dieback are classified into four stages. By stage four, the tree has more dead than live wood, the root plate has begun to shrink and the timber has become brittle, making the tree liable to fail at any time. For those who own or manage mature ash trees along roads and footpaths, they've become a punishing liability: stage four trees are no longer safe for arboriculturists to climb and must be removed using mechanical means.

The scale of the crisis affecting Europe's ash population has stimulated a comprehensive, coordinated response from plant scientists and conservationists, with Kew playing a leading role. Large stands of ash do not respond uniformly to dieback infection, and some trees display resistance, a source of optimism for the future.

The UK National Tree Seed Project has collected over 90 per cent of the genetic diversity of Britain's ash trees and safely sequestered them in Kew's Millennium Seed Bank, a resource for research and conservation. Complementing this acquisition is research on the ash genome (the totality of its DNA), which has identified the genetic basis for resistance to ash dieback – a trait thought to be held by 5 per cent of the tree's UK population. Could this trait hold the solution for the ash? Across the UK, coordinated trials monitoring resistance in ash specimens are being closely followed in the hope there is a range of reliably resistant forms that can distributed among landowners. The future of this crucial tree hangs in the balance.

15

SNAKE'S HEAD FRITILLARY

Fritillaria meleagris

The snake's head fritillary is the herald of high spring. By early May in Great Britain, the icy grip of winter is finally banished, days are lengthening to a glorious duration and the sun's warmth is tangibly nourishing. The emergence of one of our most elegant, enigmatic flowers adds to the burgeoning sense of imminent summer: a nodding flower head, composed of extraordinary chequerboard petals, that catches the sun like a glowing lantern. It grows in massed drifts, a plant wholly immersive in its beauty.

The snake's head fritillary is a plant of damp ground – and especially of that most pastoral of habitats, the river meadow. If the spring has been wet and mild, river meadows can become studded with tens of thousands of these monocotyledons, which emerge through the short grass to create a true wildflower spectacle, the equal of any alpine meadow. Sites such as Ducklington Mead, Iffley Meadows and Fox Fritillary Meadow in the UK have become sites of pilgrimage for wildflower seekers, the ephemeral nature of the fritillary's flowering only adding to the allure.

River meadows are a by-product of early land cultivation and drainage techniques that date back to the start of the thirteenth century. When the grasses were cut for hay and grazed in the early autumn, these non-exploitative management methods brought a gentle stress to the existing meadow community, ensuring no individual species dominated. Nestled among clump-forming grasses and enjoying the full sun conferred by an absence of scrub, the snake's head fritillary thrived. Many of the UK's richest habitats share a similar narrative as essentially a by-product of gentle agriculture: hazel coppices provided both a regular supply of fuel for the production of charcoal in Wealden and the perfect dappled shade environment for wildflowers and butterflies, while heathland provided grazing for ponies and became the UK's most potent habitat for reptiles.

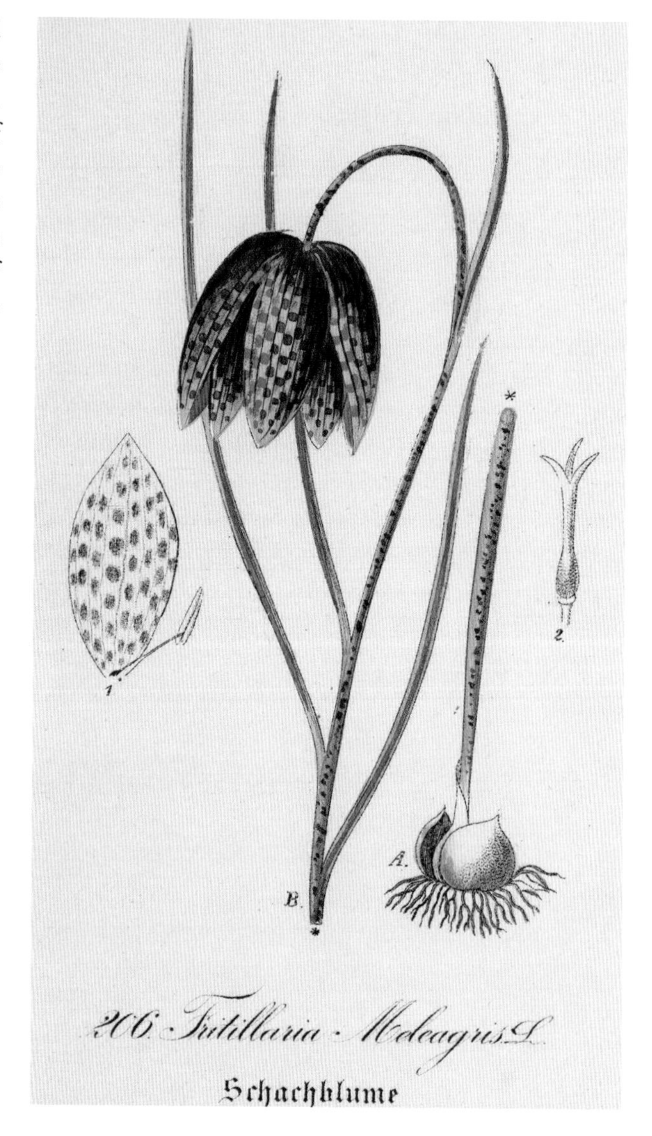

RIGHT: *Fritillaria meleagris* from James Sowerby, *English Botany*, 1869.

OPPOSITE: *Fritillaria meleagris* from Eduard Regel, *Gartenflora*, 1853. The yellow flower is *Acacia ausfeldii.*

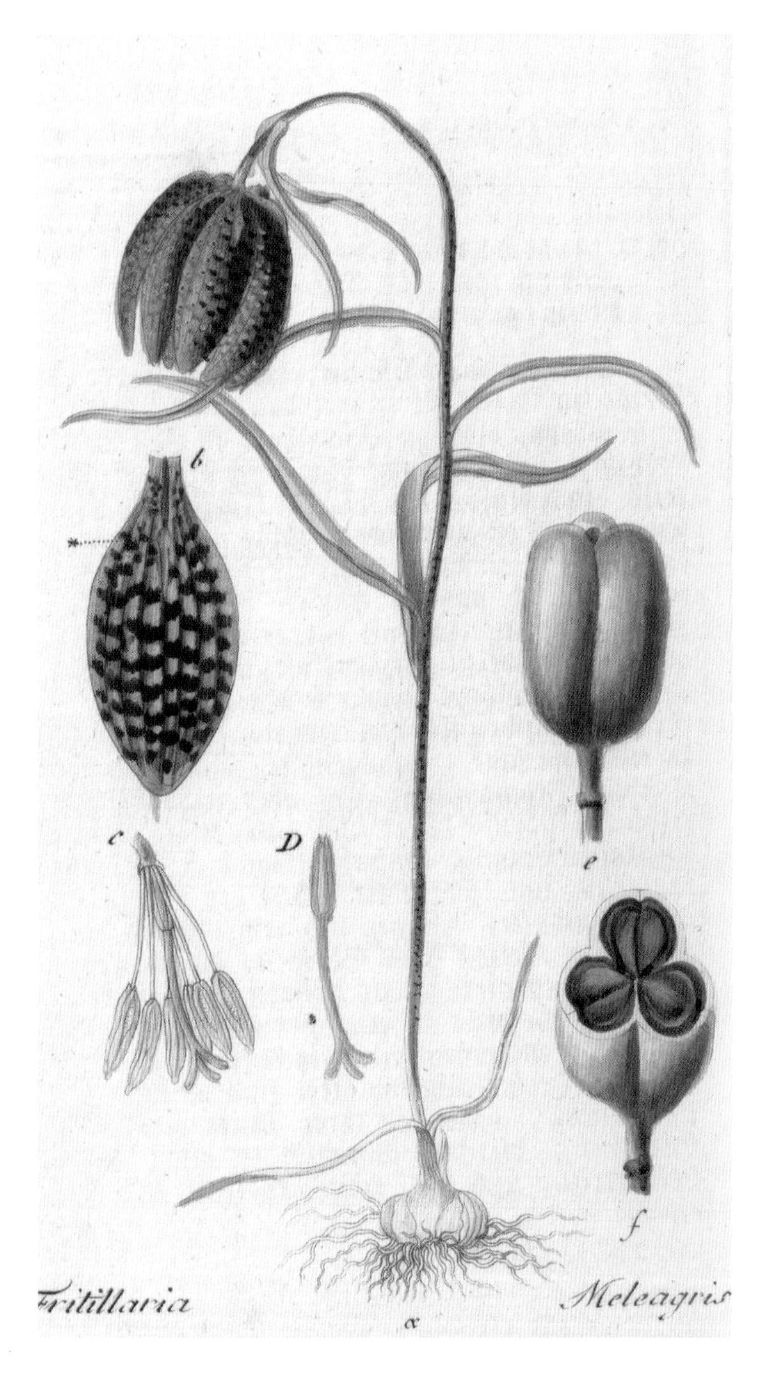

These semi-natural landscapes are wholly dependent on our stewardship to thrive – and good management is a masterful juggling of natural succession.

The gentle cycle of productivity that nurtured the snake's head fritillary underwent a dramatic shift after the Second World War. Advances in drainage allowed more intense agriculture to shift closer to rivers and the "improvement" to grassland: addition of grass-stimulating nitrogen fertilizer favoured only a select few species. Sites like Iffley Meadows saw snake's head fritillary numbers decline to hundreds of individual specimens, and it's still regarded as a scarce wildflower nationally. Climate change is also a threat to a plant delicately adapted to specific conditions. Droughts that dry the ground during spring months are as damaging as torrential rain that saturates the soil. A plant like snake's head fritillary – which needs moisture, but not too much – struggles to thrive in these extremes.

Conservation bodies such as the county Wildlife Trusts thankfully now have many of the UK's best snake's head fritillary meadows under their thoughtful management, and numbers have risen through expert conservation. Seeds of the most significant forbs (flowering plants) must have set before the summer hay is cut and the cut grass is removed to intentionally impoverish the soil. Stress and disturbance are valuable ecological commodities, maintaining pressure on the more vigorous plants and offering opportunities to those requiring space to grow and seed. Aftermath grazing – bringing sheep or cattle on the cut meadow for limited periods, the animals' feet forming scrapes and niches for seeds to germinate – does exactly this, and increases opportunities for plant diversity.

The fall and rise of the snake's head fritillary highlights our responsibility as stewards of biodiversity. Our early agricultural interventions shifted our landscape, bringing meadows, heathlands and wetlands while providing valuable materials such as hay, wood and reeds. Understanding the modern value of these habitats not just as places rich with wildlife, but also as assets capable of providing services like flood mitigation or carbon capture, drives new motivations for conservation and reasons to sustain river meadows filled with flowers.

ABOVE: *Fritillaria meleagris* from Jacob Sturm, *Deutschlands Flora*, 1804–06.

OPPOSITE: *Fritillaria meleagris* from Otto Wilhelm Thomé, *Flora von Deutschland*, 1886–89.

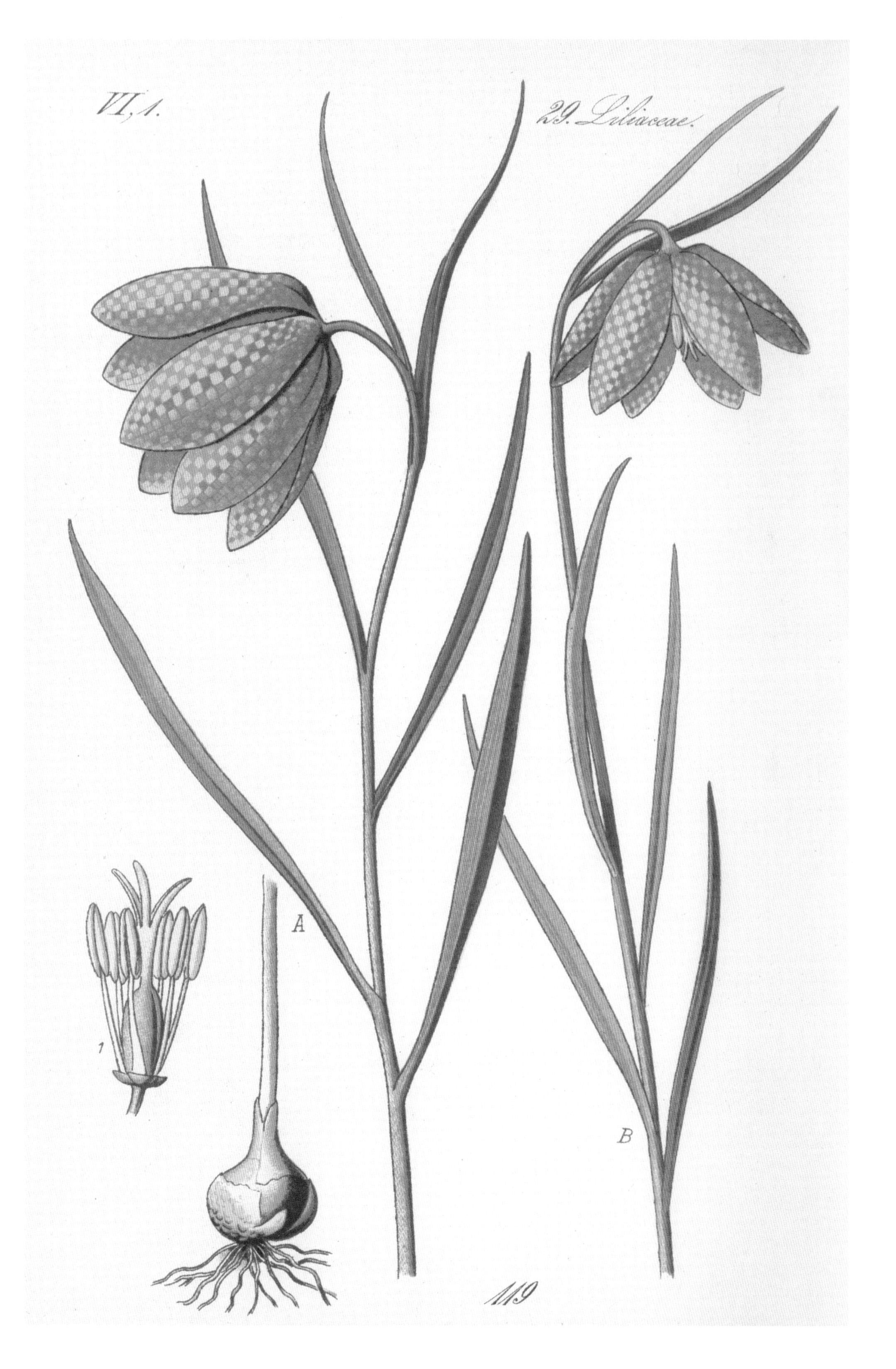
VI, 1.
29. Liliaceae.
A
1
B
119

XXXIII.
FRITILIARIA,

ALBO FLORE PVRO. PVRPVREO FLORE PVNCTATO.

MEleagris hæc planta quoque ab ave Meleagride, cuius plumas flos huius imitatur, dicta est; & Narcissus Caperonius, ab inventore, ex liliorum esse genere videtur. Huic priori cauliculus pedalis est, & interdum longior, rotundus, gracilis, firmus tamé, cavus, coloris ex purpurâ virescentis, sed obscurioris, quem sena vel plura inæquali serie ambiunt folia, brevia, angusta, nonnihil carinata; summo autem fastigio sustinet flores geminos, nutantas, nolæ vel tintinabuli instar pendulos; quorum singuli sex constant folijs, rectis, candidi coloris, circa mucronem ex viridi paululum flavescentibus; è floris vmbilico sena prodeunt stamina alba, flavis apicibus prædita, mediusque stylus itidem albus, longior, & extremâ parte trifidus, odore nullo.

Posterioris verò flos purpureus est, foris dilutioribus maculis admodum eleganter & distinctè compositis, ornatus, internè autem non minùs nigris strijs lineolisque aptè intercurrentibus, delectabilis; in cuius medio oriuntur sex staminula subflava, cum stylo eiusdem coloris. Radix ex bulbo quoque porraceo constat, subrotundo; candido.

XXXIIII.
FRITILLARIA,

FLORE LVTEO MACVLIS SANGVINEIS NOTATO. MAXIMA, PVRPVREO FLORE, POLYAN:

FRitillaria, quæ hîc prior exprimitur, florem fert sanè jucundâ colorum & macularum mixturâ insignem; cuius folia aurea, elegantibus sanguineis maculis, apto ordine, distinctis, conspersa, mirum quantùm spectantium oculos detineant; habentque insuper foris extantes per medium neruos subvirides, suam etiam flori addentes gratiam.

Altera quæ sequitur, Frittillaria maxima dicta, florum numero admodum fœcunda est; nec tamen suâ destituitur elegantiâ, cùm purpureus in singulis color, retium formam exprimere, jucundâ nunc saturitate, nunc dilutissimâ coloris differentiâ, videatur.

Pages from Crispijn van de Passe, *Hortus Floridus*, 1614.
Latin descriptions and accompanying illustrations of various *Fritillaria*.

Cirspjin van de Passe the Younger was a Dutch engraver and publisher, and member of the van de Passe printmaking family. He produced portraits and engravings for many books, however the hugely popular *Hortus Floridus* was his own project, featuring over 150 plants arranged by season, originally published in Latin and later translated into English, Dutch and French.

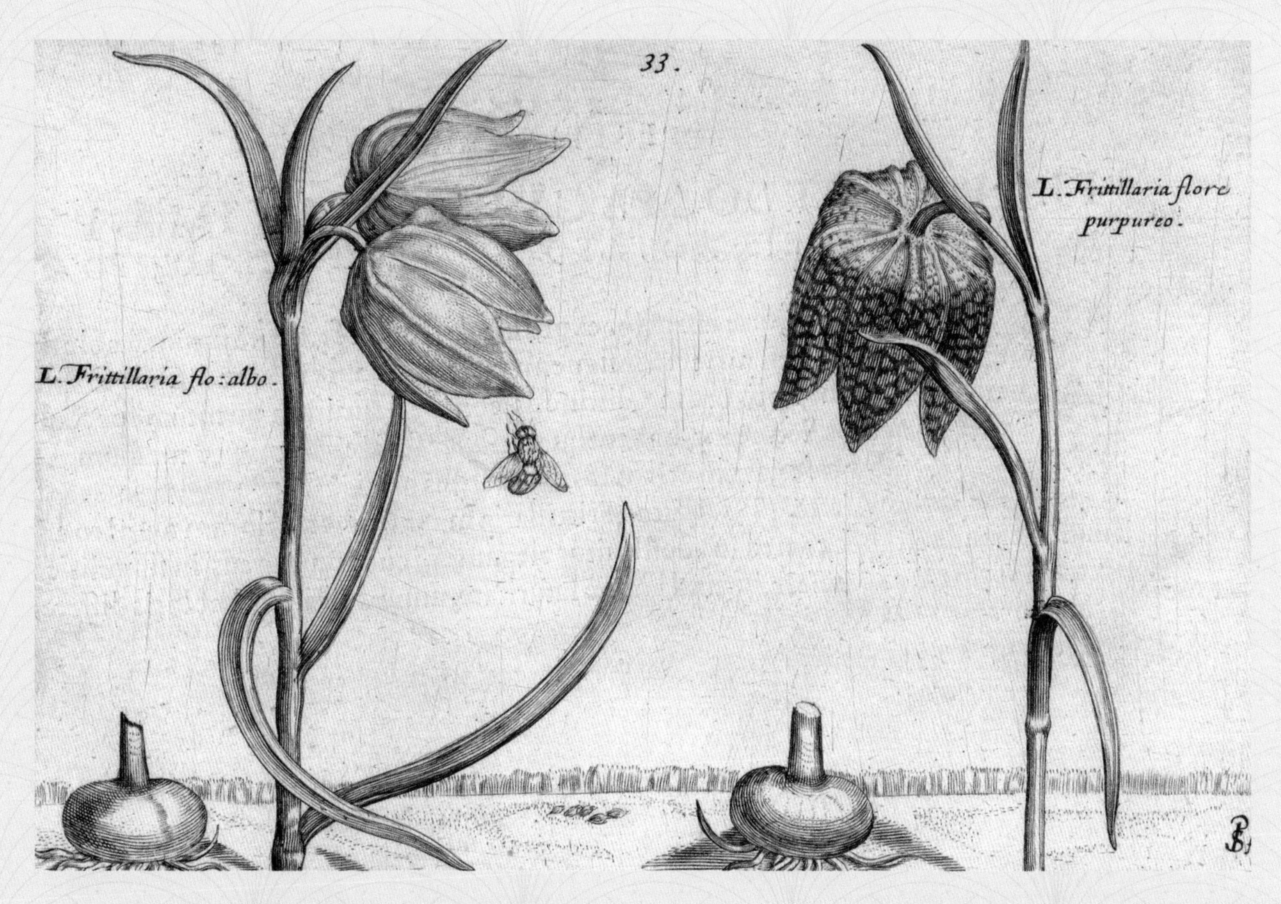
33.
L. Frittillaria flo: albo.
L. Frittillaria flore
purpureo.

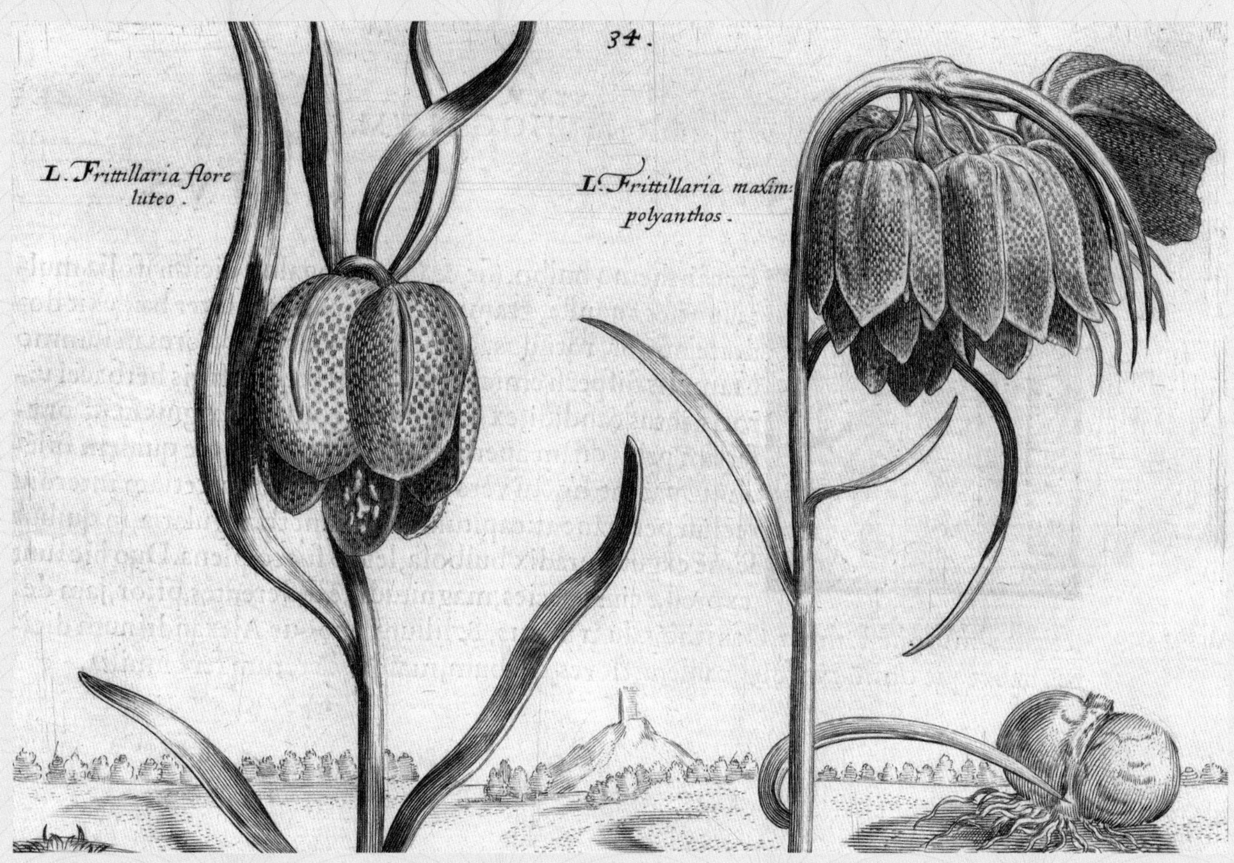
34.
L. Frittillaria flore
luteo.
L. Frittillaria maxim
polyanthos.

16

SNOWDROP

Galanthus nivalis

The snowdrop is plant that ignites passion like few others. For a few fleeting weeks every year, hordes of snowdrop fanatics, the galanthophiles, descend on gardens and woodlands to sate their obsession. *Galanthus* hotspots fill with devotees crouching low, so the pendulous flower can be lifted to observe whether a specific set of markings reveal a notable cultivar. It's not hard to see how rare snowdrop bulbs change hands for hundreds of pounds each.

The snowdrop is a plant of open deciduous woodland, an early flowering bulb that suggests winter is waning. Taking advantage of the extra light that a leafless tree canopy offers, snowdrops flower from February to April before recharging underground bulbs with an abundance of glaucous foliage. The relative paucity of airborne pollinators in February isn't an issue, with early bumblebees or even ants providing the desired service.

Galanthus nivalis is our best known and most widely grown snowdrop. Despite copious drifts found across the woodlands and hedgerows of the UK, it's naturalized rather than native to these shores, using early emergence to secure a stronghold although it's rarely aggressive or invasive. Its wild distribution is mainland Eurasia: starting in France and heading east as far as the Russian Federation. The purity of that nodding white flower (its scientific name means "milk flower") and an ability to form panoramic drifts make snowdrop displays one of the most compelling moments in the flowering calendar.

The snowdrop isn't just an ornamental commodity. The drug galantamine originates from the bulb (and is found in other related plants, such as the daffodil). Used from the 1950s onwards to arrest cognitive decline in early-onset Alzheimer's sufferers, it was initially extracted from bulbs before the active alkaloid compound was synthesized at the end of the twentieth century. Another snowdrop-derived active compound, snowdrop lectin, is an effective natural insecticide and is now being investigated for its efficacy in arresting HIV.

OPPOSITE: *Galanthus nivalis* from William Baxter, *British Phaenogamous Botany*, 1834–43.

RIGHT: *Galanthus nivalis* from Georg Christian Oeder *et al*, *Flora Danica*, 1761–1883.

230 R. DODONÆI STIRPIVM HISTORIAE

Leucoion bulboſum triphyllon.

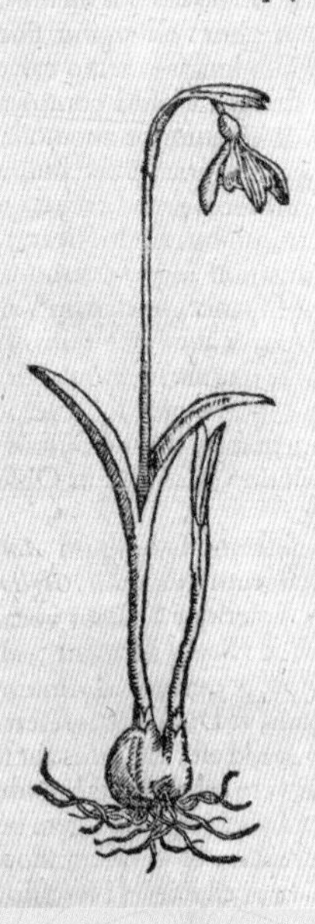

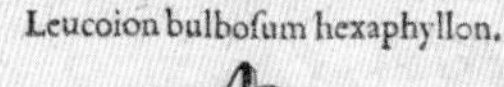

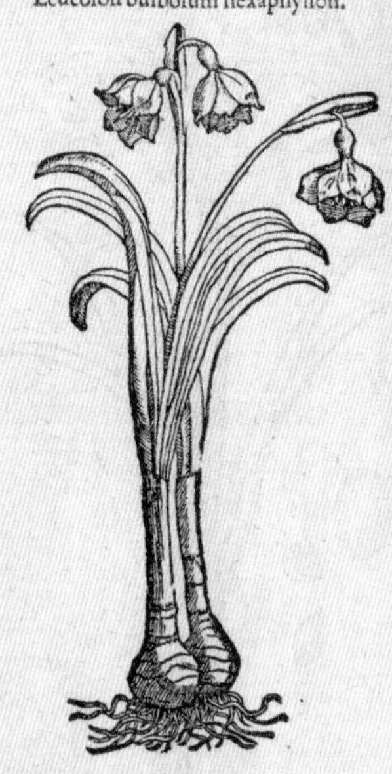

De Leucoio bulboſo. CAP. XXVI.

LEVCOII bulboſi tria nobis obſeruare genera contigit: vnum Trifolium: alterum Hexaphyllon: tertium Polyanthemon cognominari poteſt.

1. Primum folia ex ſingulis bulbis bina promit, oblonga, anguſta, in virore albicantia, foliis Pſeudo-Narciſſi lutei colore proxima, inter quæ medius exit cauliculus palmaris, in cuius ſummo flos è vaginula oblonga erumpit, à tenui pediculo deorſum dependens, foliolis tribus conſtans maioribus, candidis, intra quæ alia tria breuiora ac minora, pallidè virentia, & aliquantulùm ſtriata, in medio floris ſtamina cùm apicibus luteis: radix bulboſa eſt, foris obſcuri & eius cuius Caſtaneæ nucleus, coloris, quæ facilè in plures bulbos multiplicatur.

Peregrinus Belgio flos, ſed Italiæ & aliis regionibus familiaris.

Exit præmaturè Februario menſe, reliquos omnes ferè flores antecedens.

Leucoion videtur bulboſum Theophraſti, cuius meminit lib. 7. vbi bulbos recenſet: & quod lib. 6. primum ſeſe oſtendere ait, & emicare vbi cælum clementius eſt, ſtatim etiam nondum hieme exacta; vbi verò immitius, poſteà. Nos ad differentiam ſequentium, non inconcinnè Leucoion bulboſum triphyllon, & Violam bulboſam trifoliam, nuncupari poſſe exiſtimauimus, à tribus nempe floris foliolis.

2. Proximo generi folia ſimiliter ſunt anguſta, Porri & Pſeudo-Narciſſi ſimilia, ſed breuiora, virentiora, læuiora & magis reſplendentia, verſus terram aliquatenus reflexa: cauliculi tenues, palmares; flores in ſingulis ſinguli, rariſſimè gemini, deorſum quoque penduli, colore candidi, ex ſex foliolis aliquantulùm ſtriatis commiſſi, ſtaminibus in medio cum luteolis apicibus: ſemen in rotundis capitulis paruum, rotundum, læue, & colore flauum: radix candida & bulboſa in plures facile propagatur.

Pleriſque Germaniæ ſuperioris locis in vmbroſis & humidis ſiluis reperitur: apud Belgas hortorum areas exornat.

Floret eodem, quo ſuperius, tempore, aut non multò pòſt.

Germani Weiſſe Hornungs blumen: Noſtri Witte Tydeloosen appellant.

V Leucoij

Pages from Leonhart Fuchs, *De Historia Stirpium*, 1542. Latin description and accompanying illustrations of *Galanthus nivalis* as *Leucojum bulbosum*.

Born in Germany in 1501, Fuchs is considered one of the founding fathers of botany. His *De Historia Stirpium* was a herbal that grouped plants based on their appearance rather than on folklore and myth, with much more accurate illustrations of plants than seen before, and became the beginnings of early botanical taxonomy. The work published names in Latin, Greek and German.

OPPOSITE: *Galanthus nivalis* from Robert John Thornton, *New Illustration of the Sexual System of Carolus von Linnaeus and the Temple of Flora, or Garden of Nature*, 1807.

LEFT: *Galanthus nivalis* from Jan Kops, *Flora Batava*, 1800–46.

OPPOSITE: *Galanthus nivalis* from Amédée Masclef, *Atlas des Plantes de France, Utiles, Nuisibles et Ornementales*, 1893.

The trade in snowdrops has driven a decline in wild populations. Across the eastern side of its native distribution, *Galanthus nivalis* has been dug up for sale to the international bulb trade, in volumes that stop wild populations from regenerating. Adding to the pressure on wild snowdrops are changes in land use. The mountain foothills it favours, in regions like the Carpathian Mountains in Ukraine, are increasingly converted for housing and outdoor leisure activities. This continuous pressure on *G. nivalis* is manifested in an IUCN Red List rating of "Near Threatened", a curious contrast with the abundant naturalized displays for which the plant is famous.

The Convention on International Trade in Endangered Species (CITES) is an international accord designed to limit the decline of plants like *G. nivalis*. When these controls are successfully applied, market forces shift money to plants with a traceable, legitimate provenance and exploitatively harvested wild bulbs simply cannot be sold. The benefits from snowdrops can remain in the flower's wild countries of origin with a simple change in how the bulbs are produced.

The snowdrop is a straightforward plant to propagate. The outer "scales" that form on the bulb can be removed and induced to root, forming new plants. When raised in nursery beds, large drifts of snowdrop can be bulked up and propagated through division, lifting and splitting stocks as they increase.

By encouraging cultivation instead of exploitation and advising local growers on propagation techniques, botanic gardens can support an iconic but threatened plant. The economic value of the snowdrop is a consequence of its extraordinary beauty. Ensuring that *G. nivalis* returns to sustainable operations in its native distribution will help safeguard its future.

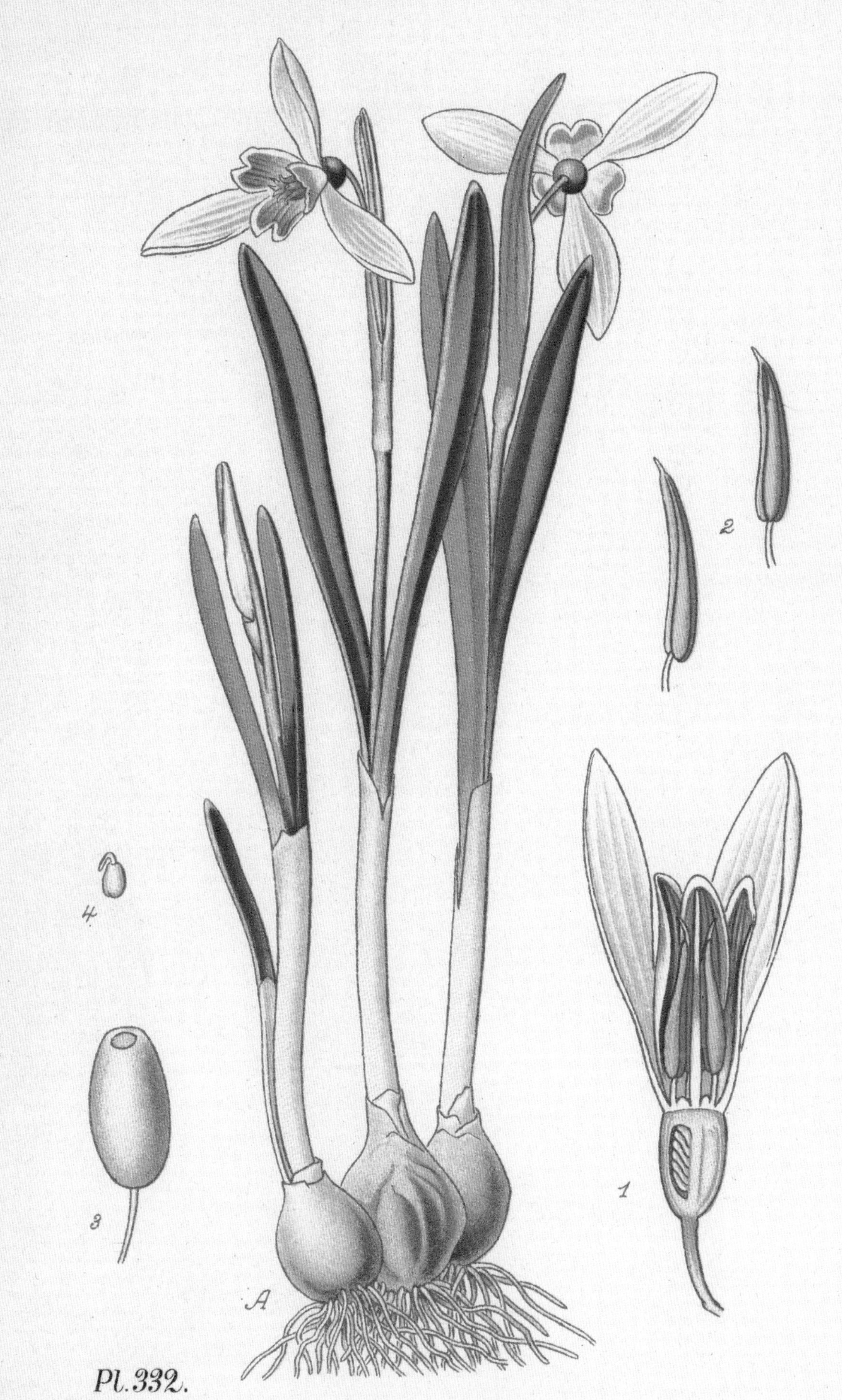

Pl. 332.

Galanthine des neiges (Perce-neige). Galanthus nivalis L.

DYER'S GREENWEED

Genista tinctoria

A beacon of high summer, Dyer's greenweed studs the UK's species-rich grasslands with spires of gold during July and August. A symbol of our early economic dependence on plants and the loss of UK meadows, Dyer's greenweed is making a comeback through enlightened conservation programmes to restore lost species diversity.

A wiry, shrubby member of the pea family (Fabaceae), Dyer's greenweed forms tight, upright mounds among the clump-forming grasses and more lax perennials in species-rich meadows. Its strikingly gold flowers are borne in a structure called a raceme and contrast most strikingly against the burnt pinks of betony (*Betonica officinalis*) and saw-wort (*Serratula tinctoria*).

Dyer's greenweed lacks the competitive vigour of pioneer meadow plants and is an indicator of late-succession, mature, species-rich grassland where species show high levels of resource sharing (known as niche complementarity). Its nectar-rich flowers are the food source for five moth species that are dependent solely on Dyer's greenweed for nutrition.

Genista tinctoria occurs in acidic, nutrient-poor grasslands (including banks and verges) across the UK, Europe, the Caucasus and central Asia. Although still present in Great Britain, it is notably absent from previous strongholds, especially in central and eastern England. It's not hard to determine the ethnobotanical origins of Dyer's greenweed and there's evidence, from Viking archaeology, of the plant being used to make dye. All parts of the plant produce a bright golden-yellow hue and can dye a range of fabrics, including leather. When mixed with the woad plant, it forms a vibrant green.

Dyer's greenweed is an ecologically uncompetitive plant and struggles to establish when direct sown by seed in meadow restorations. The first step is to reduce competition levels: cutting and collecting grass reduces nutrient levels and the addition of the notable grassland parasite plant yellow rattle further reduces vigour, creating a more level playing field. Skilled propagation methods at Wakehurst, Kew's wild botanic garden in Sussex, produce thousands of plug plants with strong root systems that are planted directly into grasslands – speeding up a succession process that would otherwise take decades.

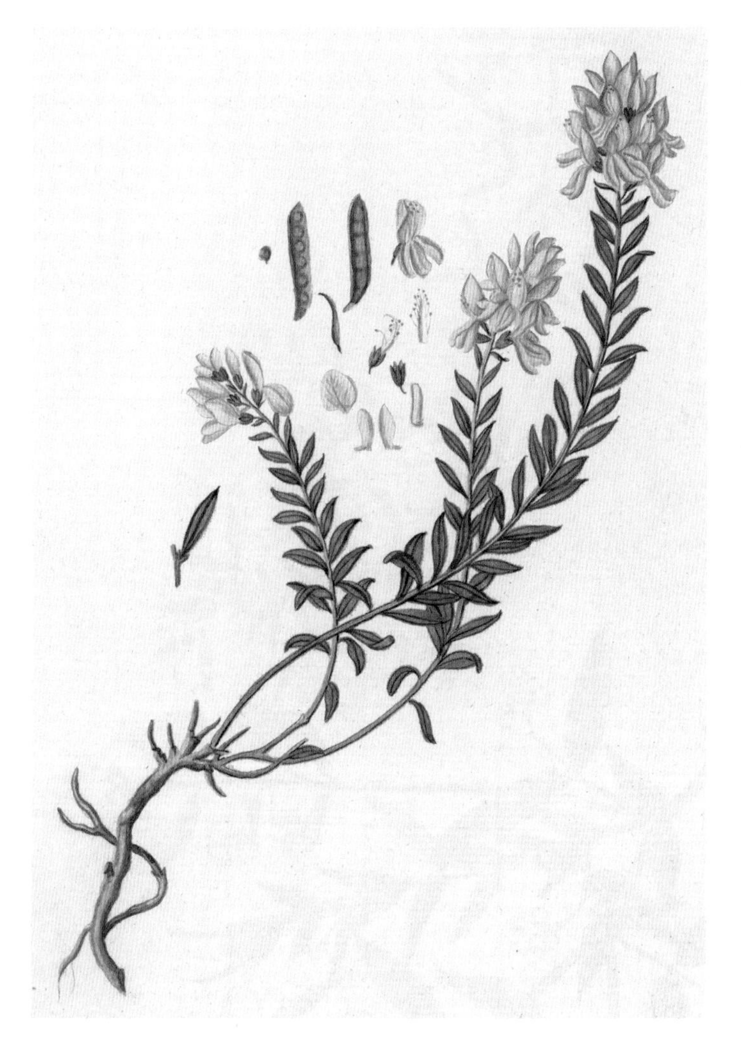

ABOVE: *Genista tinctoria* from Georg Christian Oeder *et al., Flora Danica*, 1761–1883.

OPPOSITE: *Genista tinctoria* (as *Genista virgata*) by Sarah Drake from *Edwards's Botanical Register*, 1844.

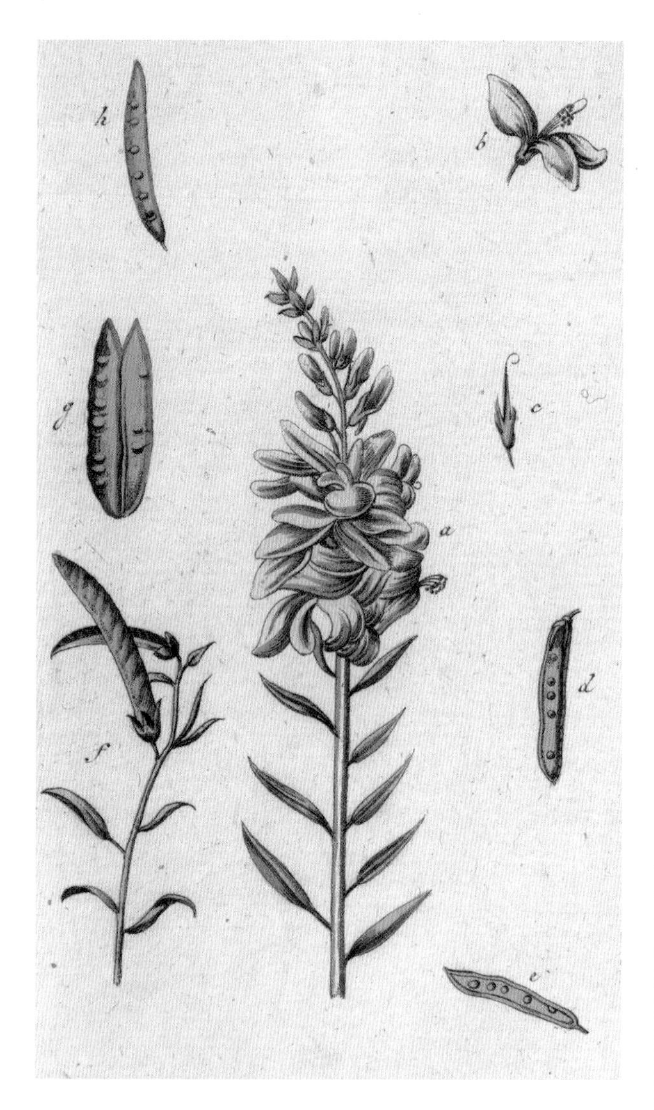

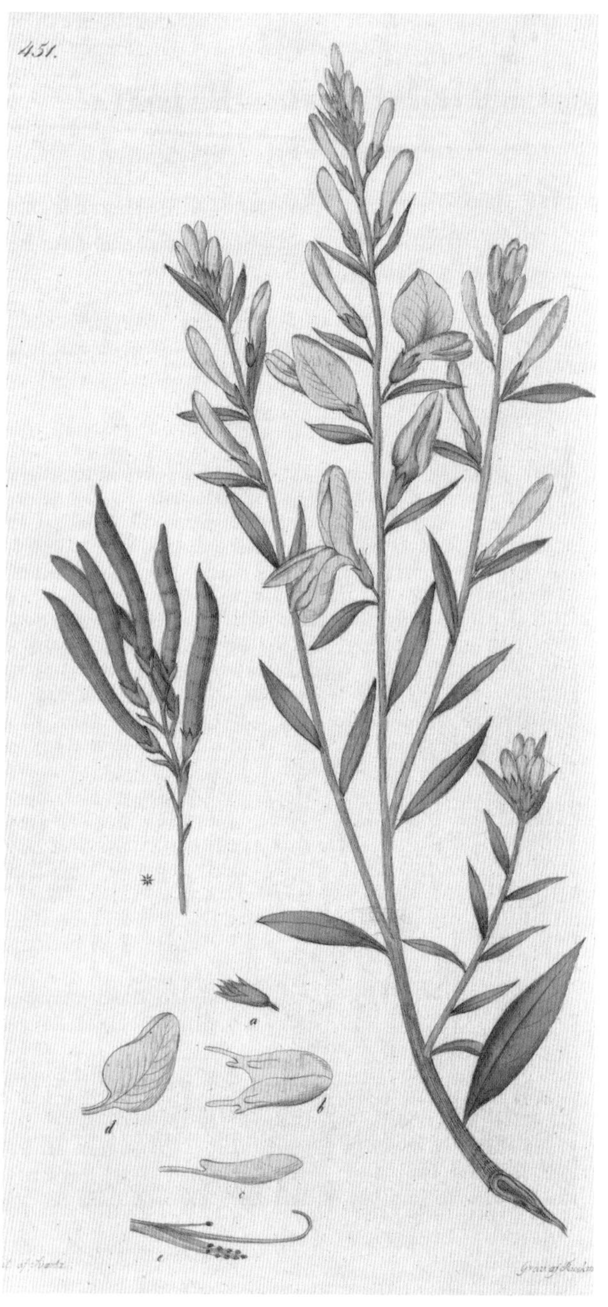

While the national British status by IUCN for Dyer's greenweed – "Least Concern" – looks reassuring, a nuanced analysis reveals a different picture. This plant, an indicator of the UK's richest grasslands, is rated "Vulnerable" in England, and with this rating comes a wider tale of dramatic meadow decline. Ninety-seven per cent of our species-rich grasslands have been lost since the Second World War, a consequence of rapidly changing land use and agricultural technologies that allow year-round field cultivation, such as winter wheat. Too often, a narrow view of land productivity means the focus is solely on the direct financial return of cultivated land.

The value of a landscape's ability to sequester carbon, provide pollinators for local agriculture or stimulate tourism and wellbeing through the power of beautiful landscape should become part of the economic equation for land management and offer fresh incentives to restore meadows, with the gleaming golden beacons of Dyer's greenweed serving as the crowning glory.

ABOVE LEFT: *Genista tinctoria* from Dieterich Leonhard Oskamp *Afbeeldingen der Artseny-Gewassen met Derzelver Nederduitsche en Latynsche Beschryvingen*, 1800.

ABOVE: *Genista tinctoria* from Johan Wilhelm Palmstruch *Svensk botanik*, 1812.

GENISTA TINCTORIA. CAP. LXIX.

Geniſta tinctoria Hiſpan.

Geniſta tinctoria vulgaris.

AD duorum cubitorum altitudinem creſcit interdum hic frutex, nudo ſtipite, enodi, recto, digitali craſſitudine, candicante cortice tecto, qui in multiplices breveſque ramos, tenellos & fragiles ſumma parte dividitur: hos ornant folia Lini aut Thymelææ, frequentia, supernè virentia, infernè verò incana & argentei planè ſplendoris, guſtus initio exſiccantis & nonnihil adſtringentis, deinde ſubamari: flores ſummis ramulis naſcuntur ſpicatim congeſti Geniſtæ tinctoriæ Germanicæ ſimiles, lutei. Tota planta elegans eſt aſpectu. *Geniſta tinctoria Hiſpanica.*

Quo tempore floret, ſub ea creſcit Hæmoderi quoddam genus elegans, pedalis altitudinis, brachialis interdum craſſitudinis, multis floribus à medio ſcapo ad ſummum uſque exornatum, magnis, oblongis, luteis, extrema parte hiantibus & in quinque partes diviſis: totum humidum eſt, pingui oleaginoſaque materia turget. *αἱμόδωρον.*

NVSQVAM hunc fruticem conſpiciebam, quàm Murciano regno ſecundùm vias naſcentem, & Martio menſe florentem. Incolæ ſtirpem ipſam *ſcopa* appellabant, Hæmoderum autem ſub eâ naſcens, *yerua tora*, quia forſitan ſi vaccæ eo veſcantur, ad venerem excitatæ taurum appetunt.

Page from Carolus Clusius, *Rariorum plantarum historia*, 1601. Description of *Genista tinctoria* with accompanying illustrations.

Another important botanical figure in the fifteenth and sixteenth centuries, and in the history of botany, Carolus Clusius published this book as a combination of his previously published books on Spanish flora and Austrian and Hungarian flora.

The "father of modern botany" Carl Linnaeus named the genus *Clusia* in his honour.

18

HIMALAYAN GENTIAN

Gentiana kurroo

Nature has few richer rewards for the intrepid mountaineer than the gentian, a sublime flower of high altitude. Alpine gentians only open in full sunshine; finding a stand in full bloom is a joyous discovery after a scrambled ascent. "Gentian" has even become the descriptive term for a rich purply blue, from the plant that has defined a colour.

The genus *Gentiana* encompasses 325 accepted species across a broad distribution, and it has some curious quirks. While those compelling high-altitude blue gentians define the genus, the large, robust and yellow *Gentiana lutea* is a surprising relative found on lower alpine meadows, while Andean gentians tend to be red. *Gentiana* is found across temperate Europe and Asia, Africa, New Zealand, Australia and the Americas. Great Britain benefits from the beauty of *Gentiana nivalis* (the Alpine gentian), which is found in the Scottish Highlands at over 900 metres above sea level.

It's a prized horticultural plant to which alpine gardeners devote time and patience, especially in replicating the flower's exacting alpine conditions: cold, well ventilated, sharply draining and sheltered from downpours. Traditional alpine glasshouses can have a quirky appearance: with a roof and no sides, they're certainly not designed to retain warmth.

Gentiana kurroo is plant of profound significance in its native Himalayan habitat. It's distributed across the north-west Himalaya, centred in Kashmir and Himachal Pradesh, and it grows at altitudes of between 1,500 and 3,000 metres. Favouring exposed southerly slopes, this low-growing perennial with richly coloured purple-blue flowers is equally happy in grassland or scree. Although beautiful, it's *G. kurroo*'s medicinal properties that define it.

As integral elements of Ayurvedic medicine, dried *G. kurroo* root and rhizomes have a wide range of applications. Used as an antiperiodic, expectorant and an astringent, and to treat chronic exhaustion, bronchial asthma and to purify the blood, this plant is a medicinal staple. It is known colloquially as "bitter root" (its scientific name derives from the Hindi word "karu", meaning bitter).

Recent research supports the use of *G. kurroo* as a medicine, and has isolated a wealth of active phytochemical compounds. Xanthones, iridoids and glucoflavones are

OPPOSITE: *Gentiana kurroo* by Harriet Thiselton-Dyer from *Curtis's Botanical Magazine*, 1880.

ABOVE: Herbarium specimen of *Gentiana kurroo* held at the Royal Botanic Gardens, Kew.

Typhonium divaricatum.—One of the most remarkable among the curious plants now in flower at Kew is this Aroid, presented to the Gardens by Dr. Regel, of St. Petersburg. It much resembles the common Cuckoo Pint of our hedges as regards foliage, but the flower is very different. The spathe is of the deepest claret colour overlaid with a velvety lustre, and the long tapering spadix is half cream and half reddish-purple. The odour emitted by the flower is by no means pleasant, reminding one of that of the Carrion Flowers (Stapelias).

Muscari Szovitzianum.—This is without doubt the finest of the Grape Hyacinths, and one that fairly represents for ordinary purposes all the kinds in cultivation, as it comes early into bloom, and continues in blossom till the latest kinds have done flowering. In colour the blooms are a beautifully clear blue, the teeth of the corolla being white. The spike is large, of an oval shape, and larger than that of other species. At Kew, the Hale Farm Nurseries, at Tottenham, and other places round London, it is among the most conspicuous of spring flowering plants.

Caltha leptosepala.—This is new to cultivation, and is quite distinct from any other on account of its flowers being white instead of yellow—a colour so prevalent in this and allied genera. The leaves resemble those of the ordinary Marsh Marigold, though somewhat smaller, and the flowers, which are nearly 1 in. across, are produced on erect stems about 6 in. high. It is an interesting bog plant, which thrives perfectly well associated with other plants of a similar character. It inhabits alpine regions in the Rocky Mountains and adjoining districts. It is now in flower in several of the hardy plant nurseries about London, and notably in that of Mr. Ware, at Tottenham.

Choisya ternata.—This beautiful Mexican shrub, of which a coloured plate was given in Vol. XII (page 232), of The Garden, will shortly be in flower against one of the open walls at Kew, where it has withstood the last three or four winters quite unprotected. Its flowers, which in general appearance resemble orange blossoms, are of a pure white waxy texture, and their perfume is also somewhat similar to that of the flowers of the Orange. Altogether, it is a first-rate shrub, evergreen, and quite hardy, at least in the south. At the Wellington Nurseries, St. John's Wood, Messrs. Henderson have some remarkably fine examples of it against a south wall.

Mertensia oblongifolia.—This little plant, which is as beautiful as it is rare, may now be seen in full flower on the rockery in the Hale Farm Nurseries, Tottenham. In general appearance it reminds one of the Virginian Cowslip (M. virginica), but it is in all respects considerably smaller. It grows about 6 in. high, and has narrow oblong leaves, covered on both surfaces with dense whitish down; and its stems are terminated by a nodding cluster of trumpet-shaped flowers, each nearly 1 in. long; some of these are prettily tinged with pink, but most of them are of a rich clear blue colour, and highly attractive. As to its hardiness there can be no doubt, as it has withstood the past winter quite unprotected; and, judging by the specimen under notice, it is also apparently both in free growth and flower.

The Double Virginian Saxifrage (S. virginiensis fl. pl).—Mr. Max Leichtlin sends us, through Messrs. Barr & Sugden, flowers of this novelty, which have just expanded in his collection at Baden-Baden. It is in every way an interesting acquisition, and moreover a pretty border or rock garden plant. The flowers much resemble those of the double Arrow-head (Sagittaria sagittifolia fl. pl.), though somewhat smaller. They form perfectly pure white rosettes, which terminate the slender branches of the flower stem. Altogether it is as interesting as the type is uninteresting; for the latter has such a weedy appearance that few would care to grow it, though no one would hesitate to give the double variety a place amongst their choicest plants, and we hope to see it in general cultivation soon.

The Rue Anemone (Thalictrum anemonoides).—This little gem might well be called the North American Wood Anemone, for it much resembles the plant which carpets our English woods in early spring. Its flowers are white but sometimes pinkish, and when seen in quantity, as at the Hale Farm Nurseries at the present time, it has a very pretty effect. It seems to thrive best in a peat border, and at the place just named it is grown in company with such plants as Mertensia virginica, Trilliums, and Lady's-slippers.

Saxifraga (Megasea) purpurascens.—We should be glad to learn from any of our readers where this handsome species can be seen well grown, especially when in flower, as we wish to make a coloured illustration of it.

National Auricula Society.—The exhibition of this society for this year will be held in the conservatory at South Kensington on Tuesday, the 20th inst., and it is expected to be of a more interesting character than usual. The recent cold weather has retarded the bloom considerably, and it is to be hoped that the northern growers will not materially suffer thereby. The new classes for Seedlings, Fancy Auriculas, Fancy Polyanthuses, and species of Primula will add greatly to the interest of the show. Roses and miscellaneous groups of plants will also be shown on the occasion, and will add to the interest of the exhibition.

THE FLOWER GARDEN.

GENTIANA ALGIDA.

This Siberian Gentian was first described and figured by Pallas in his "Flora Rossica," plate 95. It is allied to the European G. frigida, of which Grisebach regards it as a variety. In habit it closely resembles our native G. Pneumonanthe, though it is rather smaller in stature. Very robust plants are a foot high. The flowers are about 2 in. long, white, or yellowish-white, spotted and streaked with blue. Pallas found it growing with Rhododendron chrysanthum in alpine situations, and it has since been found to have a wide range. Judging from dried specimens and Pallas's coloured plate,

Gentiana algida.

this must be a very fine species. It is grown at Erfurt by Messrs. Haage & Schmidt. W. B. Hemsley.

AURICULAS IN THE OPEN GROUND.

In recommending the culture of the Auricula in the open ground, everything depends upon the stand-point taken by the writer, because there are Auriculas that would die in a few weeks if so exposed, and an abundance of others that will thrive in the open-air as well as the hardiest of border plants. It may, however, be accepted as a rule that admits of no deviation that just in proportion to the high qualities possessed by the flowers, so are the plants tender and of capricious habit, whilst the coarser the quality of bloom the hardier the plant. To trust show varieties of named sorts to the tender mercies of the weather and open-air would be madness; besides, their tender natures, their singularly delicate beauties, would be utterly spoiled by one white frost or heavy rainfall; and if these beauties are soiled they are far less effective as border plants than are the robust but coarse Continental kinds. Many seedlings from show kinds have coarse uncouth habits and quality, that render them useless for pot culture; these may be planted in the open border in preference to casting them away altogether, but none the less they are not effective border plants. I have grown Auriculas in the open ground here in stiff clayey loam, until they have become huge plants, frost doing them not the least harm; and many, having seen them, have gone into hysterics over them as being something marvellous. Perhaps it is such large showy kinds as these that excite the admiration of your Irish correspondent; but Mr. Horner is the most trust-

OPPOSITE:
Page from *The Garden*, 17 April 1880. *Gentiana algida* "a very fine species" is featured here.

ABOVE: *Gentiana kurroo* by Mary Maitland, 1823–32.

among an array of bioactive substances being proven as antimicrobial, antioxidant, antiarthritic and anti-inflammatory. The medicinal potential of *G. kurroo* is substantial, but the wild population cannot deliver this potential alone.

The Himalayas are suffering significant impact from climate change. Increased summer temperatures are pushing highly adapted species into inhabitable climate niches, the vital monsoon rains are weakening, and glaciers are receding. The precipitation that does fall is becoming increasingly unpredictable, with devastating rains destroying key *G. kurroo* habitats in Uttarakhand in northern India in 2013.

Under pressure from a changing climate and relentless wild harvesting, *G. kurroo* does not have the capacity to regenerate and thrive. Rated by IUCN as "Critically Endangered" on their Red List, the plant has seen its Indian population reduced by 80 per cent over 10 years.

Urgent conservation is required before extinction occurs and the possibility to fully research the wild population's medical potential is lost. Bioprospecting (the methodical genomic and biochemical analysis of useful plants) must highlight not just what is valuable, but also sustainable approaches to production and harvesting that doesn't impact wild populations. Time is running out for *G. kurroo*, but for plants whose usefulness is just being discovered, conservation needs to be a central outcome of any commercial interest.

MANDRINETTE

Hibiscus fragilis

Fragile by name, fragile by nature, *Hibiscus fragilis* clings to life by a thread in the wild. Its name actually refers to the brittle wood of branches, which are prone to snapping, but its wild imperilment is a curious tale of a domesticated offspring threatening its parent.

Showy *Hibiscus* cultivars bring glamour to tropical gardens. A trumpet-like flower, a bloom of extraordinary presence, occurs in arresting shades of red, orange, pink and yellow, an ostentatious stamen protruding from the petals. Bicoloured forms push the boundaries of taste, blending combinations of pink and yellow. Underpinning those eye-catching flowers is a relatively subtle evergreen shrub with toothed leaves.

The genus has a long ethnobotanical tradition, with the species *Hibiscus sabdariffa* used across China and Africa for oil, medicine and food. Recent research has isolated phytochemicals within the plant (sometimes known as "roselle") that have potential medical applications. The international trade in *Hibiscus* is significant, with a global annual production of 15,000 metric tonnes.

These plants are an international horticultural commodity: the shrubs are used extensively for landscaping in tropical, sub-tropical and Mediterranean-climate countries. Even the most ostentatious of cultivars have wild origins, with selective breeding slowly amplifying the most desirable characteristics. The journey from identifying an attractive wild trait to a commercial cultivar going on sale can take decades, a painstaking process of controlled crossing and seedling selection.

H. fragilis is elegantly understated compared to the cultivated species of *Hibiscus* – the flower's range of colour from pink to carmine red is still dramatic, but in better proportion to its glossy evergreen foliage. Experiencing its beauty in the wild is a challenge: it grows on just two mountains on the island of Mauritius. Ten known adult specimens grow on the mountain of Corps de Garde, while a relatively recent discovery on Le Morne Brabant brings an additional 26 plants to the global population. Another known population existed on the neighbouring island of Réunion but is now extinct.

ABOVE: Herbarium specimen of *Hibiscus fragilis* held at the Royal Botanic Gardens, Kew.

OPPOSITE: *Hibiscus trilobus* from Michel Étienne Descourtilz, *Flore médicale des Antilles*, 1821–9.

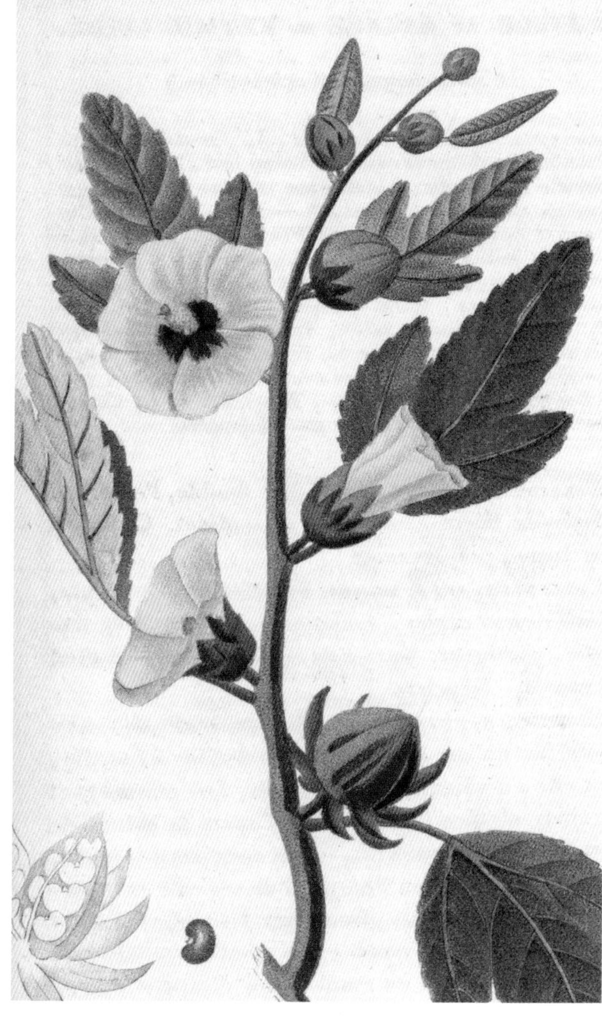

This rare plant is rated "Critically Endangered" on IUCN's Red List; the contrast between the tiny, fragmented population of the wild species and ubiquity of its robust cultivated offspring is stark. *H. fragilis* was used extensively in early breeding programmes for cultivation, the basis for a hybrid form known as *Hibiscus rosa-sinensis*, which spawned countless cultivars. Nurtured by gardeners, blessed with "hybrid vigour" and facing little ecological constraint, the garden cultivars happily adapted to new homes around the world.

In a bizarre twist, the greatest threat to the last remaining wild *H. fragilis* specimens are invasive garden plants, including the hybrid *Hibiscus*. Garden escapes can run rampant, free of the pests, diseases and environmental constraints of the wild. By colonizing ground faster – flowering or setting seed sooner – than indigenous flora, invasive species use this ecological advantage to tip the balance away from the native plants, taking light, nutrients and space from incumbents. Removal of invasive plants is rarely straightforward: roots that are not fully removed rapidly re-sprout, and stolons running on the ground or rhizomes regularly relocate the troublesome arrival.

In its original, wild form, *H. fragilis* grows in botanic gardens, including at the Royal Botanic Gardens, Kew, awaiting its moment of reintroduction. The principle of returning to the wild is a compelling narrative, but in reality, it's a complicated tale. There must be a fundamental shift in the forces that pushed a plant to extinction before its return is seriously considered. Is there a local economic benefit to that plant's return that could stimulate conservation? Has the threat of invasive plants truly been reversed? In the case of the fragile *Hibiscus*, the threat of its hybrid offspring must be dealt with first.

ABOVE LEFT: *Flowers of Roselle*, by Marianne North, 1870.

ABOVE RIGHT: *Hibiscus sabdariffa* from Michel Étienne Descourtilz, *Flore médicale des Antilles*, 1821–9.

OPPOSITE: *Hibiscus indicus* (as *Hibiscus venustus*) by Matilda Smith from *Curtis's Botanical Magazine*, 1891.

7183
2.
4.
3
1.
M.S. del, J.N Fitch lith.
Vincent Brooks, Day & Son Imp.

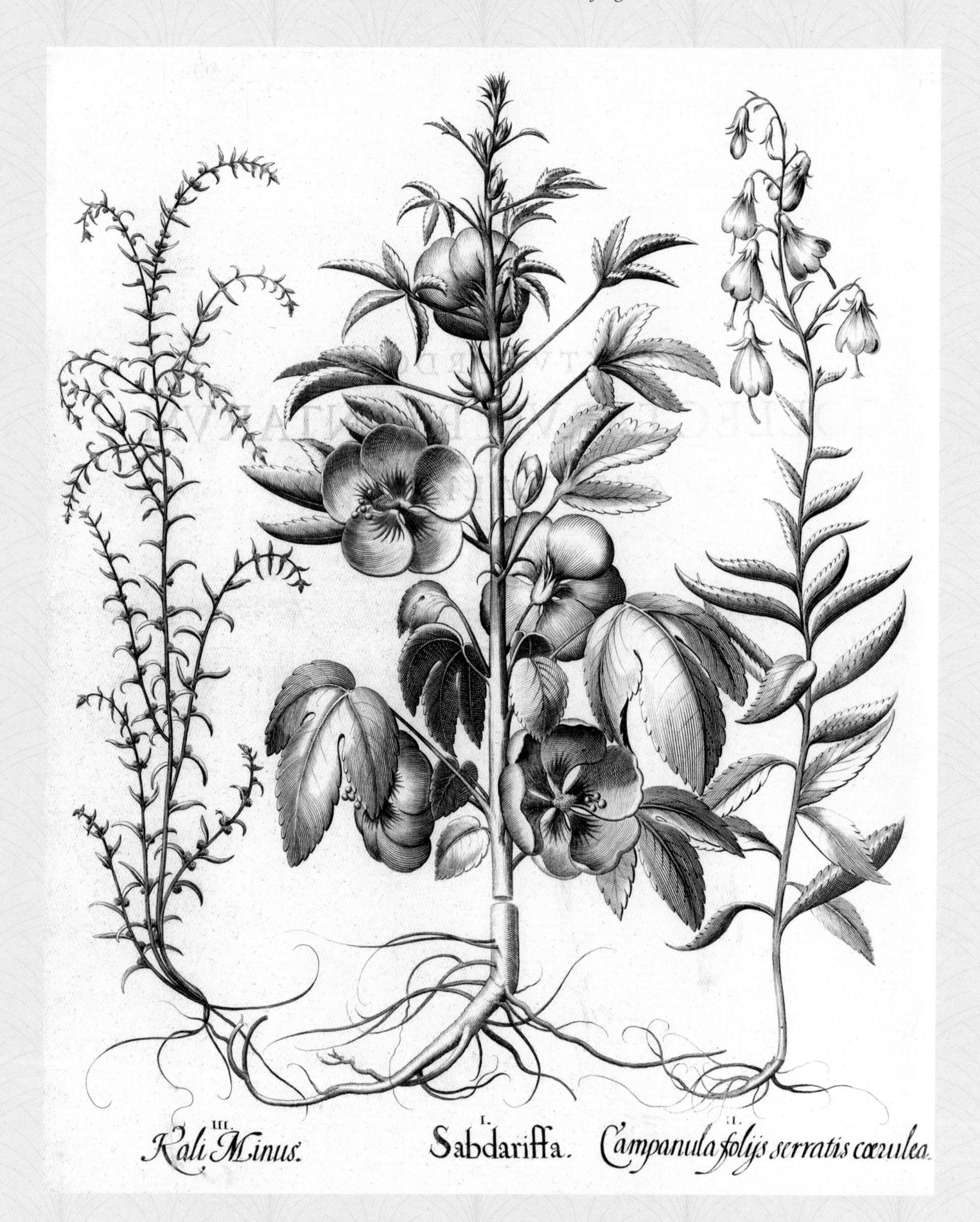

Hibiscus sabdariffa **from Basilius Besler,** ***Hortus Eystettensis*****, 1613.**

Famously known for his *Hortus Eystettensis*, Besler was a German botanist and apothecary. His book was a record of the plants growing in the garden of, and commissioned by, the Bishop of Eichstätt Johann Konrad von Gemmingen. Much like his contemporary Crispijn van de Passe in his *Hortus Floridus* (see page 92), the book was arranged by season, reflecting the life of the garden at Eichstätt.

RIGHT: *Hibiscus cameronii* by Walter Hood Fitch from *Curtis's Botanical Magazine*, 1842.

20

THE TUNBRIDGE FILMY FERN

Hymenophyllum tunbrigense

This implausible fern appears from a distance as a glossy green film coating the sandstone outcrops it inhabits. Closer inspection reveals distinct flattened fronds, pressed tight to the nurturing humidity of its geological foundations. The filmy fern highlights the fragile dependencies some plants form with their environment.

Unlike higher-flowering (angiosperm) plants, which have efficiently incorporated reproduction and growth into one neat structure, ferns have a separate "gametophyte" form to produce offspring, a life stage that is completely reliant on moisture.

Hymenophyllum tunbrigense is, like other filmy ferns, monocellular with fronds (the fern equivalent of leaves) one cell thick. The fronds are broad and flat and held just above the surface they grow on by wiry rhizoids (primitive fern roots). Mature populations of filmy ferns cover whole sandstone cliff faces.

Ferns are ancient plants that first evolved when the Earth was a continuous land mass. This leads to some extraordinary distributions among fern species, with the filmy fern a notable example. In addition to being native to Europe, it's also found in a selection of African countries (including Kenya, South Africa and the islands of Madagascar and Mauritius) as well as superficially random outliers such as Jamaica, New Zealand and South Carolina.

In Great Britain, it's primarily found where it's mild and wet: the west coasts of England and Wales as well as in Ireland. But there's a curious population in the south-east, where the extraordinary sand rock habitats of the High Weald simulate the steady Atlantic fret of west Cornwall.

Previous attempts to grow the filmy fern in horticulture have emphasized its profound dependency on highly specific environmental factors. Sustained primarily by atmospheric humidity in the wild, the filmy fern requires twice-daily

OPPOSITE: *Trichomanes speciosum* (as *Trichomanes radicans*) from Thomas Moore *The ferns of Great Britain and Ireland*, 1855. *Trichomanes* is a relative of *Hymenophyllum*, both in the Hymenophyllaceae family.

RIGHT: *Trichomanes speciosum* (as *Trichomanes radicans*) from Edward Joseph Lowe *Ferns: British and exotic*, 1825–1900.

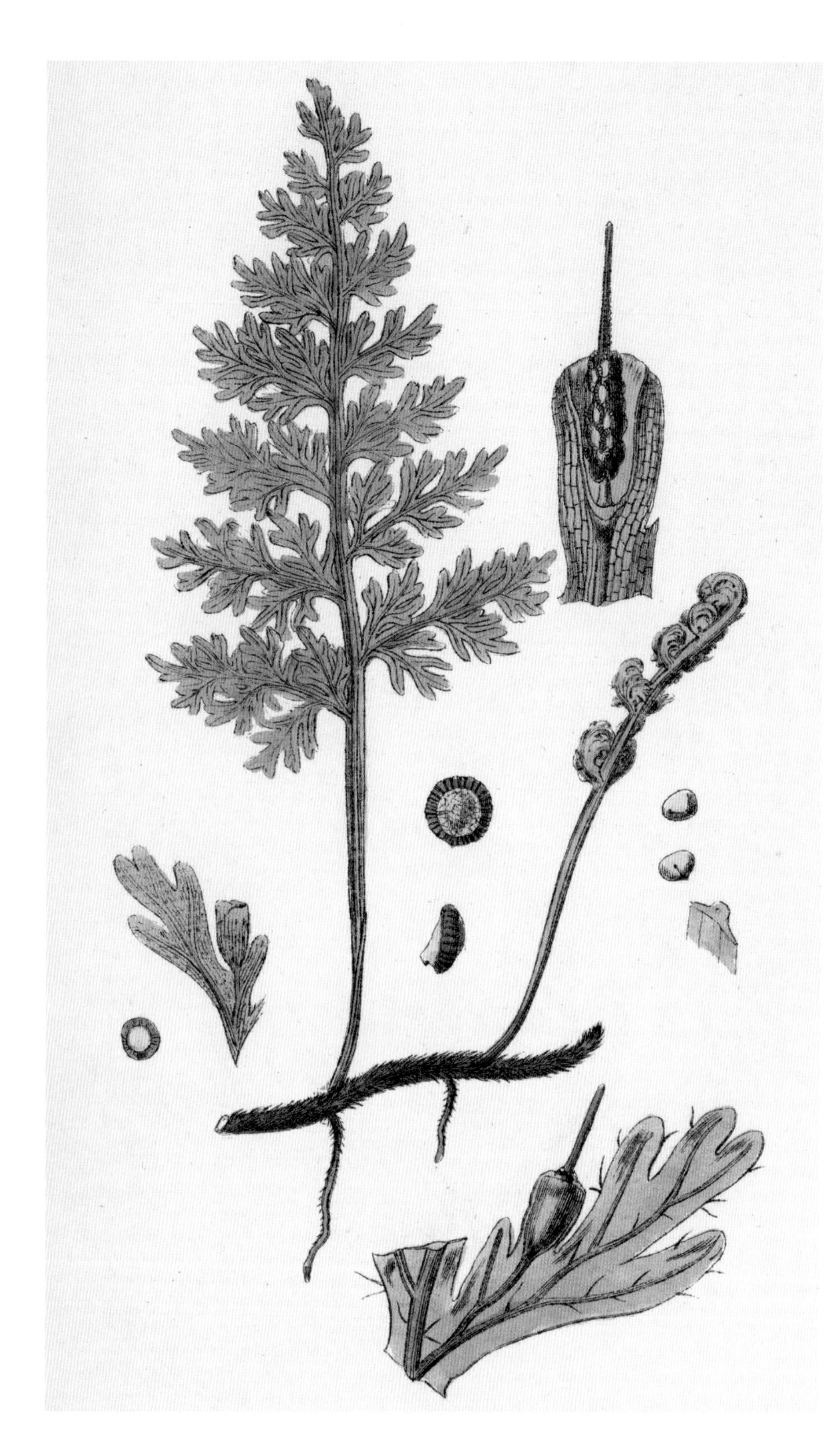

LEFT: *Trichomanes speciosum* (as *Trichomanes radicans*) from James Sowerby, *English Botany*, 1886.

misting with mineral-free water in horticultural glasshouses. The challenges of growing the filmy fern in cultivation make it hard to comprehend why anyone would want to steal it from the wild for a private collection; the theft of an exceptional High Weald population from National Trust property Nymans in West Sussex, UK, in 2011 prompted bewilderment as much as anger.

The need to live in near-constant humidity makes life in increasingly difficult for the filmy fern, and 72 per cent of recorded filmy fern populations were lost between 1950 and 1995. Extended periods of drought combined with high temperatures push the resilience of a plant with fronds one cell thick to the limit. The filmy fern is heavily dependent on the dappled shade conferred by open beech and oak canopies. Woodland clearance spells the end for a plant unable to adapt to direct sunlight.

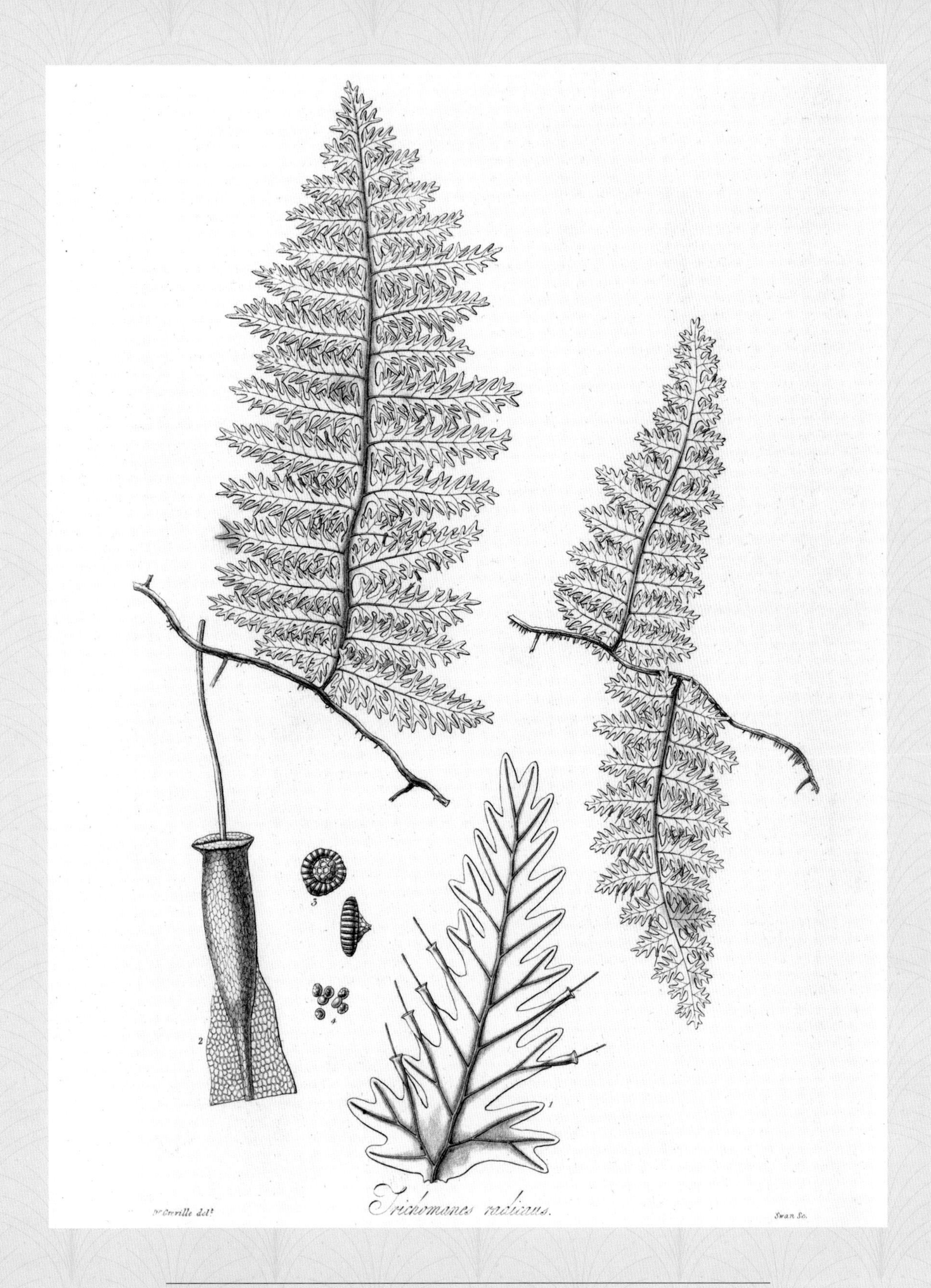

***Trichomanes speciosum* (as *Trichomanes radicans*) from Sir William Jackson Hooker and Robert Kaye Greville, *Icones filicum*, 1831.**

Published before Hooker became Director of Kew (1841–65), *Icones filicum* was one of several publications he produced during his professorship of botany at the University of Glasgow. His co-author Greville was a botanist and mycologist, as well as the illustrator for the book.

21

SOFAR IRIS

Iris sofarana

If humanity is the biggest threat to the natural world, isn't the answer simply to declare our most biodiverse habitats protected reserves and shut people out? Some conservation models separate people and nature, so they occupy distinct and rarely overlapping territories. Human residence and economic activity happen in one space, and nature in the other. Whether reality can ever be this discrete is a hot topic among ecologists.

Cities are not clinical environments but a range of novel niches filled by opportunistic species, including tree of heaven, coypu, terrapins and peregrine falcons. What appears to be the most pristine wilderness may have been shaped by millennia of human stewardship. Should a new model for conservation accept how intertwined humans and nature truly are, and steer us towards a middle ground that benefits both?

Lebanon has significant biodiversity, notable endemism and high population density. At the junction between three continents, with dramatically diverse landscapes and a nurturing Mediterranean climate, this country of only 10,000 square kilometres has over 2,000 plant species, including 108 endemics, and a population of six million people. Development pressure is high, with habitable space concentrated by mountainous terrain.

Iris sofarana is a captivating resident of Lebanon's highlands. Endemic to the western slopes of Mount Lebanon between 1,200 and 2,000 metres, this species has further variations in a distinct subspecies called *kasruwana*. It's a striking plant, as dramatic in colour and form as any nursery-bred cultivar. Rising 20–30 centimetres out of its preferred grassland habitat, the flowers are broad and bold, the lower petals a marbled bronze, and the upper petals veined in fine streaks of purple. It's not surprising to learn this elaborate colouration has evolved to attract a single specific pollinator, a solitary long-horned bee (solitary bees don't live in a hive, produce honey or serve a queen bee). Outside the iris's flowering period, the bee seeks refuge in hollowed logs or rocky crevices near its nectar source.

With a limited distribution, burgeoning pressure from human population and reliance on a sole pollinator, the Sofar iris has slim chances for survival. Encroaching developments

LEFT: *Iris sofarana* by Pierre Joseph Redouté from Pierre Joseph Redouté, *Les Liliacées*, 1802–6.

OPPOSITE: *Iris* plants from Johann Wilhelm Weinmann, *Phytanthoza iconographia*, 1737.

N. 611.
d
a
b
c
a. Iris Anglica bulbosa flore purpureo.
b. Iris Susianna.
c. Iris bulbosa variegata.
d. Iris vulgaris humilis flore violaceo.

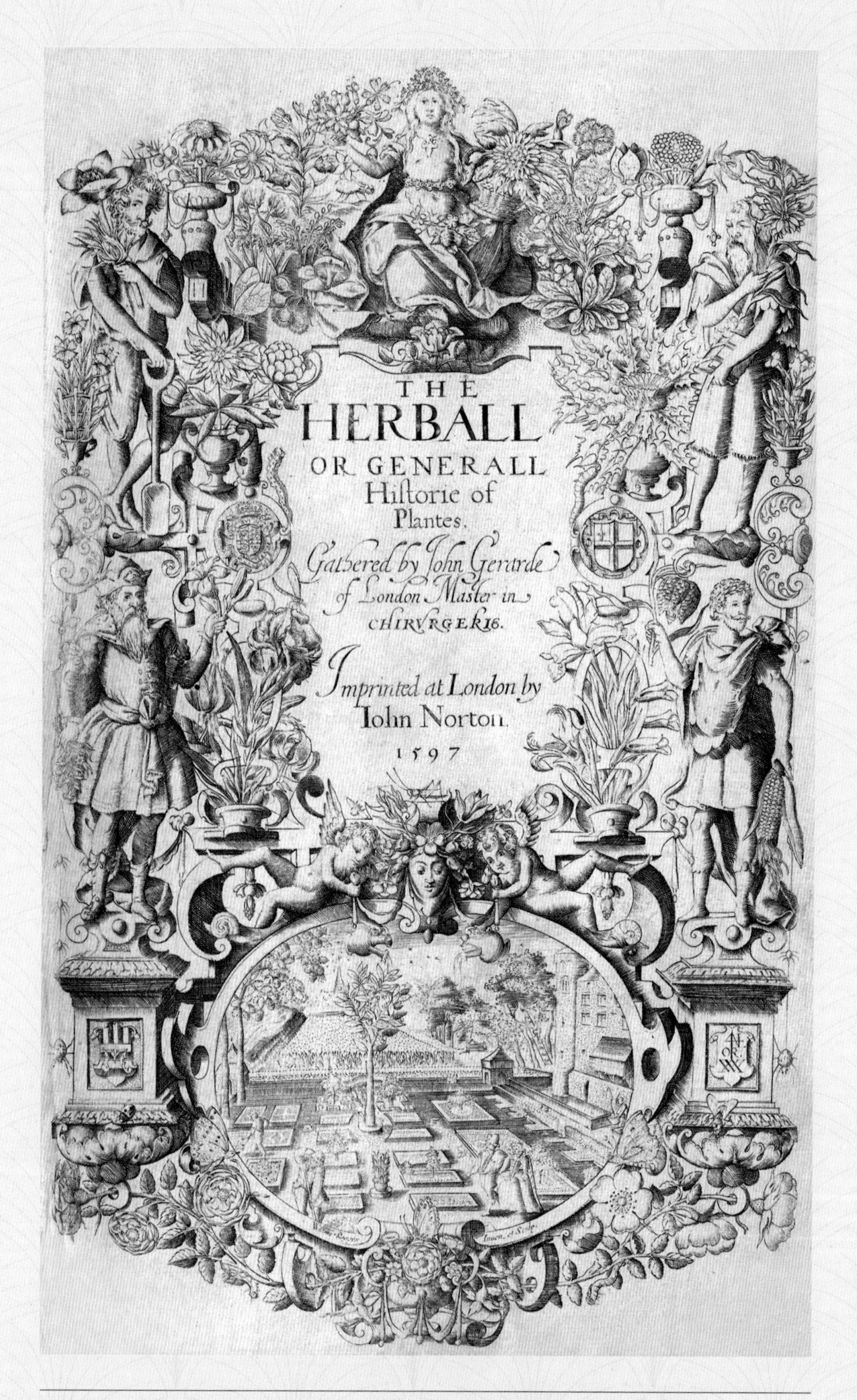
THE
HERBALL
OR GENERALL
Historie of
Plantes.
Gathered by John Gerarde
of London Master in
CHIRVRGERIE.
Imprinted at London by
Iohn Norton.
1597

Title page featuring a small garden with many popular flowers including *Iris sofarana*, from John Gerard, *The Herball* or *Generall historie of plantes*, 1597.

Gerard's *Herball* was an English translation of Dodoens' work (see page 165) with additions from his own garden in Holborn, London, and plants from the New World. A later edition was published in 1630 after Gerard's death, with additions and corrections by apothecary Thomas Johnson.

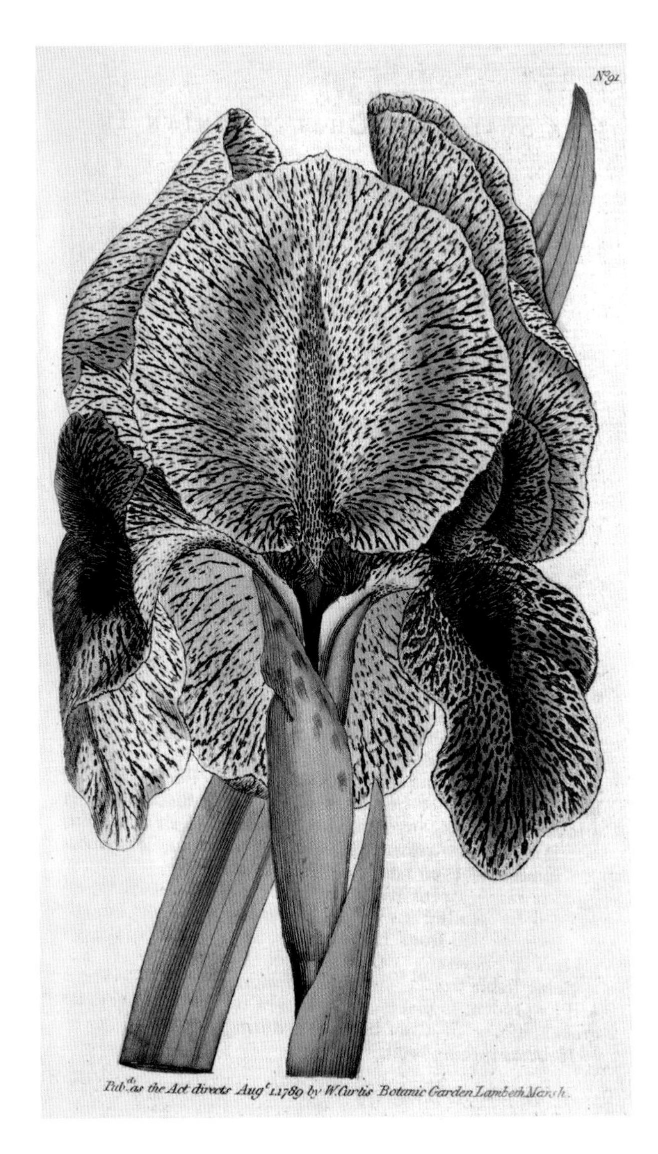

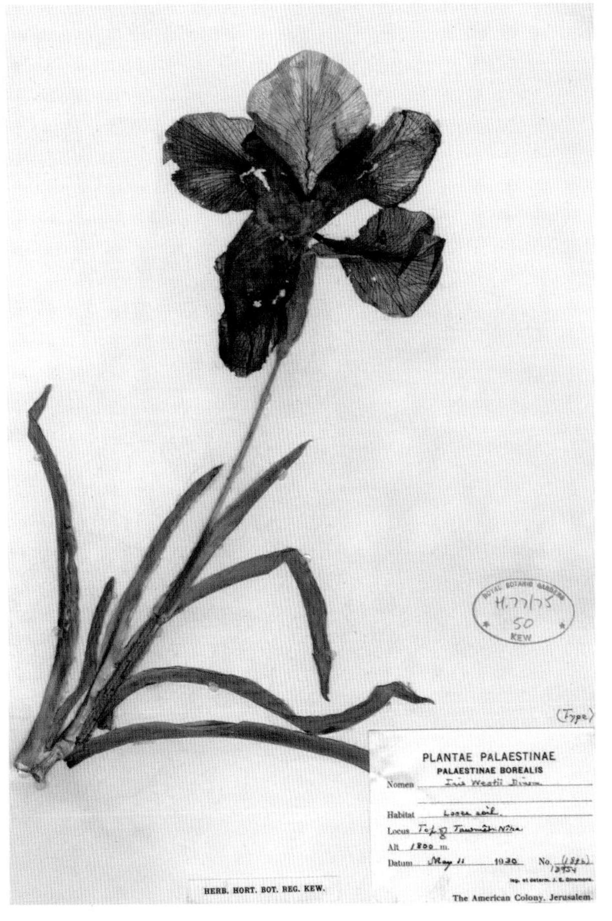

LEFT: *Iris sofarana* from *Curtis's Botanical Magazine*, 1790.

ABOVE: Herbarium specimen of *Iris sofarana* collected by J. E. Dinsmore in Palestine in 1930, held at the Royal Botanic Gardens, Kew.

such as ski chalets and removal of the solitary bees' habitat through pesticide use are threatening the plant's future, and it is categorized as "Endangered" on the IUCN Red List. Is this another collision course between human and nature, destined to end with the latter critically endangered?

Lebanon is pioneering a new approach to land management – a balance between conserving biodiversity and provisioning human need – and the results are promising. The country's Shouf Biosphere Reserve is a place for people *and* nature, with an innovative accounting system showing just how valuable that nature is. By analyzing the ecosystem services that the flora and fauna of the reserve provide to people, a compelling ratio emerged. For every US dollar invested in conserving biodiversity, 19 are returned.

This analysis uses sophisticated modelling to show how nature benefits us. Healthy ecosystems filter water (reducing how much processing is required before drinking), offer pollinators for our agriculture, sequester carbon, provide food and support our cultural history. Beautiful, diverse landscapes are also highly desirable places to stay. Hotels and guest houses in the Shouf Biosphere Reserve can charge significantly more than their peri-urban equivalents, such is the premium placed on a view of nature. With Shouf exemplifying a new approach to conservation, there's hope this approach will spread to other sites, with potential to save *I. sofarana*.

Nature provides, regulates, supports and serves as a source of culture for humans; these are desirable services with quantifiable values. A shifting economic model, reflecting our inextricable shared connection, starts new conversations about where to spend funds. Investing in nature could be the answer to protecting biodiversity and concurrently improving our livelihoods.

JACARANDA

Jacaranda mimosifolia

The jacaranda tree is a plant inextricably linked with warmth and sunshine, a sensory connection to warm climes, the memory of a holiday to the Mediterranean. When jacarandas bloom in late spring or early summer, they give the impression of being more flower than tree, a profuse cloud of purple. Often used as street trees, they transform their environment, bringing immersive beauty to their urban surroundings while tolerating the heat, pollution and poor soil that comes with the territory. Their slender form adds to their compatibility with an urban environment.

Beyond the joyful beauty of their flowering, the story of the jacaranda becomes more complicated. Is it a threatened tree, endemic to a limited range of South American forest? An iconic species with cultural resonance in South Africa, Australia, California and China? Or is it an invasive weed that requires a licence to be planted? It is, in fact, all of the above.

The contrast between the jacaranda's wild and cultivated distributions is extraordinary. Its native range is limited to the piedmont forest of north-west Argentina and southern Bolivia. This dry montane forest is highly biodiverse, containing at least 80 endemic species of both flora and fauna, including the military macaw, and is rapidly being cleared for agriculture. The crop replacing this rich habitat is primarily soybeans, a commodity driven by global demand for emulsifiers (most chocolate bars are bound together by soy lecithin) and animal feed.

The remaining piedmont forest has become fragmented, with small disconnected "islands" of primary habitat. Species that have evolved to flourish in continuous forest cover may struggle in a fragmented habitat: plants may be unable to seed in forest edge or open ground, and animals may be easily predated away from the densest forest cover. The threat to piedmont forests has led to an IUCN Red List rating of "Vulnerable" for the *Jacaranda mimosifolia*.

Away from the constraints of the piedmont forest, where they are nurtured and not cleared, the species lives a parallel life to its wild ancestors. The South African administrative capital of Pretoria is known as "Jacaranda City", defined by its abundance of heavily flowering street trees. The plant's willingness to adapt to the warmer temperate cities of the

OPPOSITE: *Jacaranda mimosifolia* from *Revue horticole*, 1897.

ABOVE: *Jacaranda mimosifolia* from Louis van Houtte, *Flore des serres et des jardin de l'Europe*, 1847.

HERB. HORT. BOT. REG. KEW.
CENTRO DE PESQUISAS FLORESTAIS E CONSERVAÇÃO DA NATUREZA
TIJUCA - RIO DE JANEIRO
EX-HERB.
N.º 1804.
Jacaranda mimosifolia D.Don.
Guanabara, Avenida Édison Passos.
Leg. C. Angeli 3 XII 1962 nº 332.
Det. J. Corrêa Gomes 1963.
Obs. Árvore de ± 8 m. Flôres roxas.

world has nudged it into invasiveness. It's illegal to sell *J. mimosifolia* in the South African nursery trade, and new public plantings require a licence.

The factors causing a species to become invasive are both environmental and ecological. In native distribution, a whole suite of constraints regulates the growth and reproductive capacity of the plant, maintaining its integration within a community. Climate, soil, pests, diseases, parasites or a dependence on specific pollinators or mycorrhizal symbionts are all factors controlling vigour.

Placed in a novel context with fewer or no constraints, some plants change function and become dominant and pioneering. Trees kept within highly stable woodland in their native distribution can become weedy when disturbed ground is available to colonize.

A fascinating and ongoing debate concerns the very nature of indigeneity and whether the notion of "native" and "exotic" are more of a human construct than we'd like to admit. Some scientists point to natural hybridization and migration (by air, water or animal vector) and suggest human activity is simply accelerating a natural process. The house sparrow is, for example, essentially an introduced species that has thrived in human-disturbed habitats, but we accept it due to its endearing nature. Other avian interlopers, such as the ring-necked parakeet, are vilified due to a perceived coarseness.

J. mimosifolia – the tree that's slowly disappearing in its native land while overrunning its new surroundings – highlights this nuanced debate. Should we welcome thriving exotic trees that are capable of providing shade, beauty and soil stabilization in tough conditions, and suspend our ecological concerns? The only certainty is that the issues the jacaranda highlights will only continue to challenge us.

OPPOSITE: Herbarium specimen of *Jacaranda mimosifolia* collected by C. Angeli in Brazil in 1962, held at the Royal Botanic Gardens, Kew.

ABOVE: *Jacaranda tomentosa* (as *Jacaranda jasminoides*) by M. Hart from *Botanical Register*, 1827.

23

CHILEAN WINE PALM

Jubaea chilensis

While many associate palm trees with tropical forests or an idyllic island paradise, the Chilean wine palm grows in the temperate climes of Chile, tolerating cool winters and dry summers as the most southerly representative of its family. The palms (Arecaceae) are a family of considerable economic importance, supplying global trade with rattan wicker, coconuts, dates, sugar, fibres, wax and oil. The date palm is no less valuable, and its edible mini coconut-like seeds, delicious sap and basket-making leaves leave it prone to over-exploitation.

As our climate changes, horticulturists search for new caulescent (trunk-forming) plants to complement or supersede the trees traditionally chosen for future environmental niches. Palms are increasingly used to bring height and structure to urban gardens, and the Chilean wine palm has proven reliably hardy in London, thriving in hot, dry summers and cold winters (it can tolerate temperatures down to -15° Celsius).

ABOVE: *Jubaea chilensis* (as *Jubaea spectabilis*) from Oswald Charles Eugène Marie Ghislain de Kerchove de Denterghem, *Les Palmiers*, 1878.

OPPOSITE: *Jubaea chilensis* from Karl Friedrich Philipp von Martius, *Historia Naturalis Palmarum*, 1823–50. The genera *Prestoea* (as *Martinezia*) and *Latania* are also depicted here.

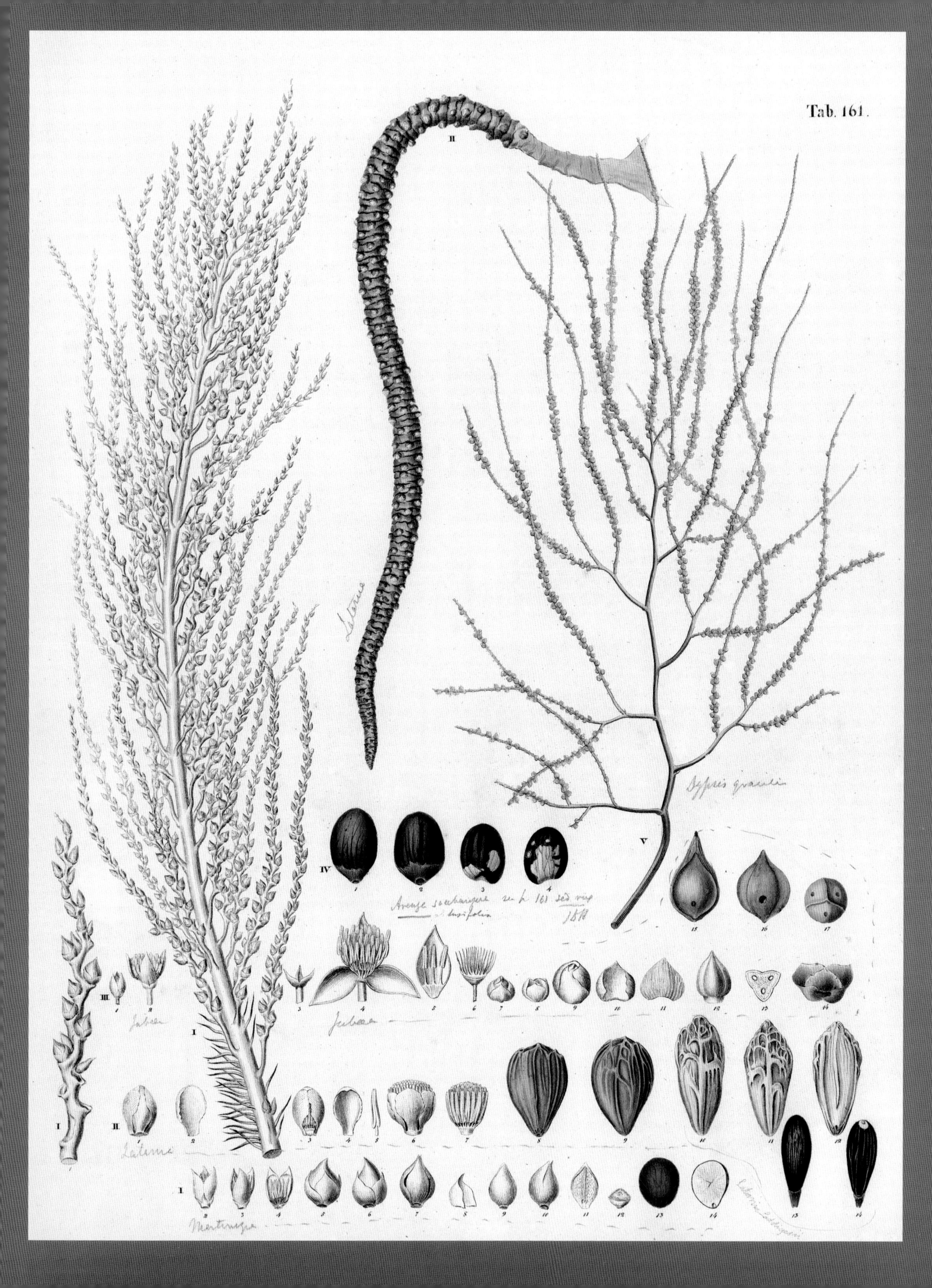
Tab. 161.
I
II
III
IV
V

ABOVE: *Specimens of the Coquito Palm of Chile, in Camden Park, New South Wales*, by Marianne North, 1880.

OPPOSITE: *Jubaea chilensis* (as *Jubaea spectabilis*) from Eduard Regel, *Gartenflora*, 1853.

Jubaea chilensis is a dramatic plant with impressive dimensions. A broad trunk, punctuated with serrated, diamond-shaped leaf scars, slowly grows outward to become the thickest of any palm tree, ranging between 1 to 1.7 metres at maturity. The trunk is an ever-expanding single organ of living tissue – unlike a tree trunk, which develops girth through annually added rings of non-living wood.

Mature Chilean wine palms are imposing specimens, clocking in at over 30 metres tall. They're a compelling addition to any landscape, but those who desire instant horticultural impact may be frustrated by their slow-growing nature. This trait, coupled with an ability to survive stressful transplantation, means Chilean wine palms have often been dug up from the wild for the nursery trade.

It took a long time for the Chilean wine palm at the Royal Botanic Gardens, Kew, to reach maturity. Raised from seed in 1846 and planted in the nurturing environment of the Temperate House, it slowly rose towards the firmaments of the magnificent glasshouse, earning the sobriquet "the world's largest houseplant" along the way. Finally, an inevitable and unsustainable stand-off between growing palm and roof occurred, requiring a fresh chapter of Chilean wine-palm growing at Kew.

Back in the wine palm's native Chile, its abundance of natural capital was, for many years, its undoing. The sugary sap is extracted and sold as palm syrup or fermented into palm wine. The manufacturing of palm wine is common across the palm family's distribution, but while many species can be sustainably "tapped" to draw sap from a living plant, accessing the wine palm's most desirable product is a more destructive process.

To harvest syrup, wine palms are decapitated, causing up to 400 litres of sap to vigorously pour out of the terminal wound. With no capacity to regenerate its growing tip, this is the palm's final act. Wild populations have thus been swiftly reduced and fragmented, leading to an IUCN Red List rating of "Vulnerable".

Protection under Chilean law appears to be slowing the rate of decline, but legal imposition is rarely the complete solution. Sustainable or less exploitative use of natural resources can create a more equal sharing of benefits between biodiverse environments and people. A renewable, economically viable product for the user may discourage intensification of land use while providing long-term stability for the plant community and the life it supports. Finding new methods to extract syrup from the Chilean wine palm without ending the plant's life is the smart solution to safeguarding wild populations of this extraordinary plant.

371

ARNOLD ARBORETUM, HARVARD UNIVERSITY.

Ansd. 1.V.06

Jamaica Plain, Mass., April 18, 1906.

ROYAL GARDENS 30 APR. 1906 KEW

Dear Colonel Prain:

I find here on my return your letter of January last containing a list of a few plants that you want for the Kew Arboretum. It is too late to send them this spring but you shall have as many as we can send in the autumn, as well as several other things of which Bean made a memorandum the other day.

I brought home from ~~Peru~~ Chili a few cans of the prepared sap of Jubaea spectabilis which is used in Chili as a sort of treacle under the name of meal de palma. I shall be glad to send you one of these if you care for it for the Kew Museum.

I was kept late in the Garden the other afternoon with Elwes and Miss Willmott, and when I called with the latter at your office to see you you had gone. Am very sorry to lose the opportunity to say good-bye. Let me know always if there is anything that we can do for you on this side of the ocean.

Faithfully yours,

C. S. Sargent.

Col. D. Prain,
Royal Bot. Gardens, Kew,
London, England.

Letters from Charles Sprague Sargent, Arnold Arboretum, Harvard University, Jamaica Plain, Massachusetts, USA, to Sir David Prain, Director of Kew, 18 April 1906 and 15 May 1906.

Sargent writes to Prain to say that he has collected a few cans of the prepared sap of *Jubaea spectabilis* (*Jubaea chilensis*) which is known locally in Chile as "miel de Palma", a treacle-type substance. He says he will be glad to send a sample if desired – today this is held in Kew's Economic Botany Collection.

He writes to Prain again on 15 May 1906 to confirm that a can of this prepared sap is being sent via American Express.

372

ARNOLD ARBORETUM, HARVARD UNIVERSITY,
JAMAICA PLAIN,
MASS.

ROYAL GARDENS 26 MAY. 1906 KEW

Ackd. 11.VI.06.

May 15, 1906.

Dear Colonel Prain:

I have your note of May 1st and we are sending you by the American Express Company a small can of the Jubaea spectabilis juice for the Museum. I am sorry it weighs too much to go by mail.

Faithfully yours,

C. S. Sargent.

Colonel D. Prain,
R. Bot. Garden,
Kew, London.

Thanks herewith.
53/1956 June 9
J.H.H.

OPPOSITE: *Chilean Palms in the Valley of Salto*, by Marianne North, 1880.

TANGLE WEED KELP

Laminaria hyperborea

In the chilly, turbid inland waters of the Sussex coast in the UK is a habitat of extraordinary biodiversity, one that regulates the waves and absorbs vast volumes of carbon dioxide – and which we've treated with disregard. This habitat is the wondrous underwater forests of kelp.

Those lucky enough to dive among our kelp forests enter a surreal, otherworldly kingdom. Perpetually swaying to the sea's motion, tall "trunks" (kelp stems are called stipes) create a rhythmic environment accentuated by flitting shoals of fish seeking the forests' shelter. Translucent fronds filter light onto a host of other seaweeds, diverse in form from feathery to filamentous. Exploring the kelp forest may reveal cuttlefish, seahorses, crabs, sharks or even a seal. This is the temperate analogue for coral reefs: complex, biodiverse and a vital environmental buffer between sea and land.

The term "kelp" is broad, referring to the order Laminariales, although this high taxonomic rank is populated by only 30 genera. The kelp forests of Sussex are primarily composed of three species: tangle weed (*Laminaria hyperborea*), oarweed (*Laminaria digitata*) and sugar kelp (*Saccharina latissima*). These common names suggest that the latter species is more valuable than the former two, and this is in fact the case. Kelp has an exceptionally long evolutionary history, dating back to over 20 million years ago, in the Miocene era.

Kelp's ability to anchor to the sea floor (using an adaptation called a holdfast) is the foundation for its role as habitat manager and environmental engineer. Because kelp is a biological constant in an ever-changing environment, a wide diversity of species flock to the stability of the kelp forests, making it a nursery for many fish, crustaceans and molluscs. This richness of life offers an abundant larder for predators, a remarkable food web (over a thousand species have been recorded in the UK's kelp forests) underpinned by the plant's exceptionally efficient photosynthesis.

This photosynthetic power is borne out in a range of dizzying statistics. Kelp forests can absorb 20 times more carbon dioxide per acre than their terrestrial equivalent, and it's estimated the world's kelp forests absorb 600 million metric tonnes of carbon dioxide annually. It's the method

OPPOSITE: Red algae *Gracilaria* by Ellen Hutchins, early nineteenth century. Ellen Hutchins is considered Ireland's first female botanist, who specialised in seaweeds, lichens, mosses and liverworts.

ABOVE: *Laminaria cloustoni* from Köhler, *Medizinal-Pflanzen*, 1887.

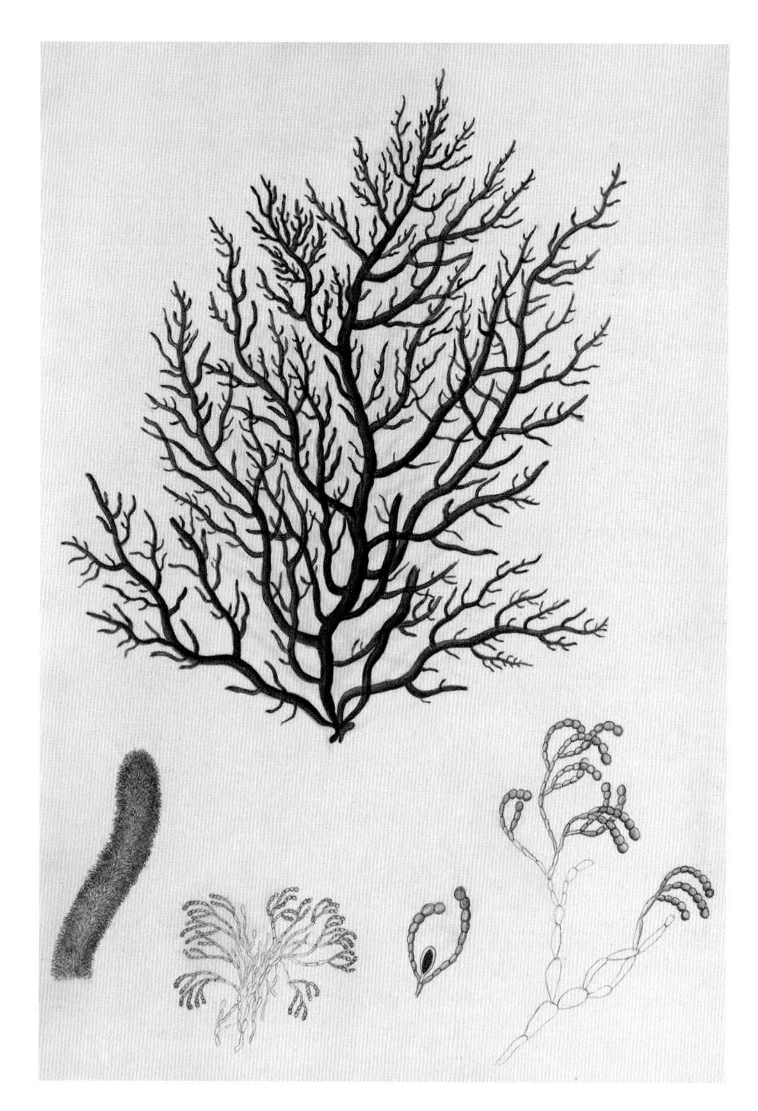

ABOVE AND LEFT: Seaweed by Ellen Hutchins, early nineteenth century.

of absorption that's significant, too. Growing rapidly, deep underwater, the carbon dioxide absorbed by kelp is sequestered deep into the ocean, safely stored away from the atmosphere.

It's not just its prodigious powers of carbon capture that make kelp valuable to us. The thick swaying forests have the power to slow and regulate sea motion, removing energy from large waves and reducing on-shore impact. The value of kelp to nature and people should surely have driven best practice in conservation, with large protected zones ensuring this biological wonder thrived. Unbelievably, quite the opposite happened.

Progress in fishing technology, with new bottom-dredging trawlers scraping up significant volumes of ocean life, increased efficiency but dealt a massive blow to our kelp forests. The holdfasts could not hold fast against intensified fishing, and mature kelp plants were swept away, with regeneration impossible against constant harvesting. Further threat came from the practice of "losing" sediment, pumped at high pressure over inshore waters. This undesirable substance is alien to plants adapted to clear waters, that prodigious photosynthesis impossible where there's no light. Over 95 per cent of the kelp forests of Sussex, UK, have been lost.

Slowly, a compelling body of research, evidencing both the value of kelp forests and the alarming extent of their decline, has been assembled to effect change. The Sussex Inshore Fisheries and Conservation Authority is in consultation with local authorities, the fishing community and conservationists to implement a trawling exclusion zone, giving kelp vital space to regenerate. Must it take a species declining by 95 per cent to mobilize action? The lesson of the vital, valuable plant we pushed to the brink must be learnt.

261 318

Antermony,
Glasgow, 17th Feby
1841.

My Dear Sir,

My father has desired me to say that he received your letter of the 12th Ulto: and at the time being unwell he delayed answering it from day to day; and at present he has been laid up by an attack of pneumonia, which altho' it has been overcome by the usual remedies of bleeding and low diet, has left him in rather a languid state.— We can sincerely sympathise in the family distress which you and Lady Hooker have undergone; and trust that your next communication may bring us more favourable accounts than your last letter.— Eliza Horrocks continues an invalid; and although at present better than I ever expected to see her, the prospect of her complete restoration to health does not appear to be near at hand.— I do not know of any publication which contains a full, or detailed account of the manufacture of, and trade in, Kelp.— In Ure's Dictionary of arts there is a short account of it, and rather a more full one in a small pamphlet which I printed about a year ago on the Neapolitan Sulphur question; and of which I shall endeavour to send you a copy.— Iodine is still prepared from drift weed Kelp in the course of the process for seperating the alkaline salts from the Kelp. This process you will find also described in Ure's Dictionary. You are probably aware that a committee has been ordered by the house of Commons to enquire into the practicability of relieving the persons ruined in the Highland Kelp trade, by means of emigration at the expense of the country. In the Times, (London paper), of the 13th Feby you will find a good deal of statement which may interest you in relation to the Kelp trade in the speech of Mr Baillie the Member for Inverness, in moving for the said Committee. More than two years ago I warned the Government that this expensive measure would become indispensible if they persisted in their policy in regard to the Neapolitan Sulphur question; and I repeated the warning in my printed pamphlet.— All, however, was as I expected, wholly unheeded by our legislators; and I cannot say I expect any thing like sound policy, good feeling, or even common sense from the proceedings of the Committee of Parliament upon this subject. The evidence taken before the Committee, will, however, most probably be printed; and ought to form a useful & interesting source of information to those interested in the history, & situation

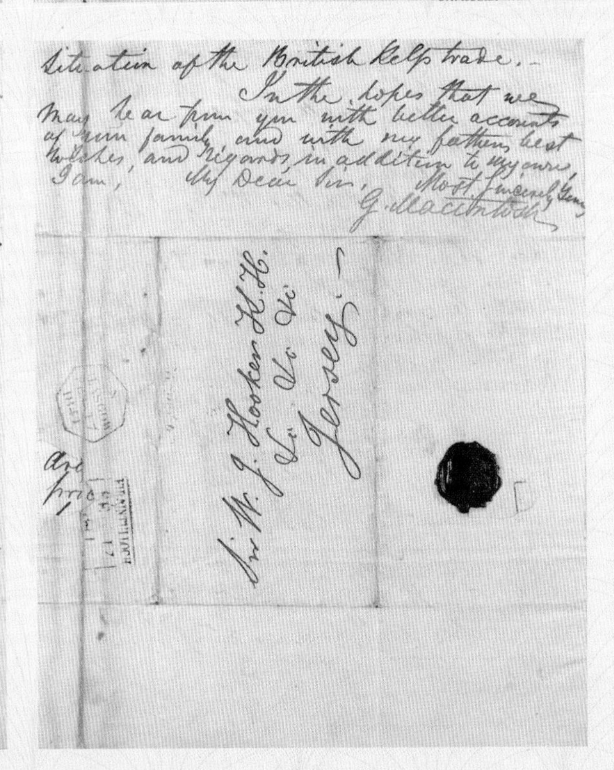

situation of the British Kelp trade.— In the hopes that we may hear from you with better accounts of your family, and with my father's best wishes, and regards, in addition to my own, I am, My Dear Sir, Most sincerely Yours,
G. Macintosh

Sir W. J. Hooker K.H.
&c &c &c
Jersey

Letter from George Macintosh, Antermony, Glasgow, Scotland, to Sir William Jackson Hooker, 17 February 1841.

Macintosh writes to Hooker regarding the British kelp trade and the House of Commons committee that has recently been set up to address the situation in the Highland kelp trade. Macintosh has no faith in the legislators and does not expect any sound policy from them, but concludes that evidence from the committee will serve as a useful history and record of the kelp trade.

PICO DE EL SAUZAL

Lotus maculatus

Studded with waxy orange-red flowers reminiscent of exotic birds' beaks, and boasting tumbling glaucous foliage, the alluring *Lotus maculatus* is the subtle star of many botanic garden glasshouses. One of four *Lotus* species endemic to the Canary Islands, *L. maculatus*'s wild distribution occupies a mere square kilometre on Tenerife, and it is unsurprisingly classified as "Critically Endangered" on IUCN's Red List. Its future is dependent on urgent conservation action.

The Canary Islands are a hotspot for biodiversity: over 500 endemic species occur over the archipelago's 7,500 square kilometres. Formed by volcanic eruption and featuring sharply varied topography and an equitable climate, these islands scattered off the west African coast have perfect conditions for speciation. Extraordinary plant genera are unique to the islands: succulent *Aeonium*, towering *Echium* and the fine-foliaged daisy *Argyranthemum*.

Lotus maculatus is a seaside dweller, tolerant of salt spray and an integral component of low-growing coastal scrub, alongside samphire, *Scilla* and statice. Wild displays of *L. maculatus*, tumbling over undulating clifftop terrains, were a spectacular sight, but the plant is now rarely seen – only 35 known individuals are left. Though it is restricted only to the Paisaje Protegido de la Costa de Acentejo reserve,

BELOW: *View of the Peak of Teneriffe*, by Marianne North, 1885. Tenerife cliff top view, similar to where *L. maculatus* might have grown in abundance in the past.

OPPOSITE: *Lotus maculatus* by Christabel King from *Curtis's Botanical Magazine*, 2008.

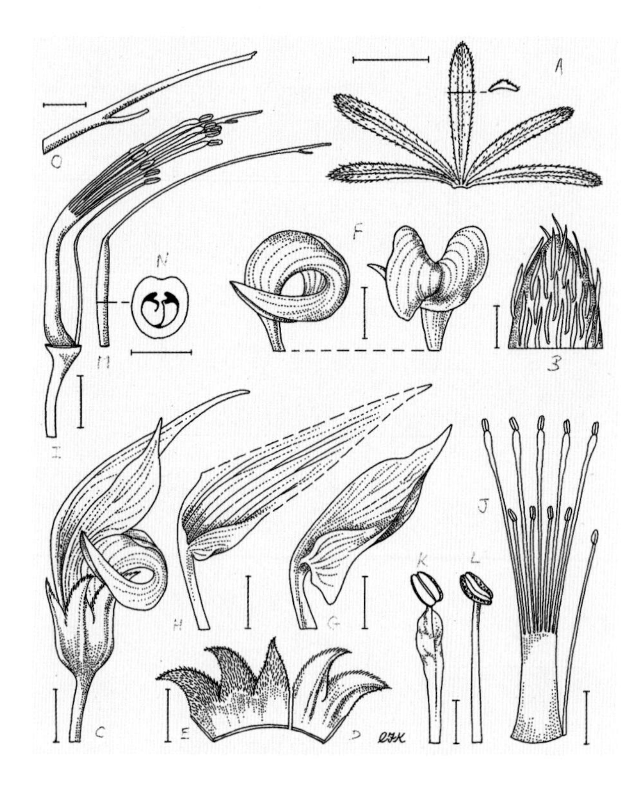

ABOVE: Line drawings of dissections of *Lotus maculatus* by Christabel King from *Curtis's Botanical Magazine*, 2008. This was published along with the painting on page 137.

RIGHT: *Lotus peliorhynchus* (as *Lotus berthelotii*) from *Revue horticole*, 1876.

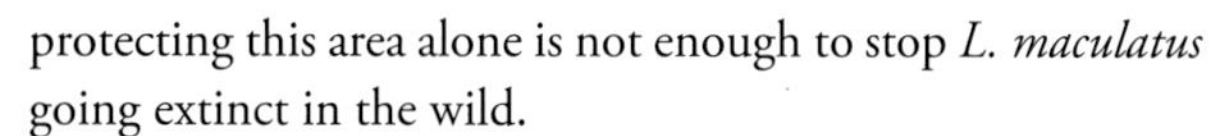

protecting this area alone is not enough to stop *L. maculatus* going extinct in the wild.

Threatened by introduced grazing mammals (especially rabbits), trampling, and changes in land use, *L. maculatus* became fragmented and diminished to a perilous state. While grazing pressure has been successfully reduced in the Paisaje Protegido reserve, fundamental changes in ecology are harder to mitigate against.

An increased seagull presence (linked to human activity) has led to higher volumes of guano, richer soil and an invitation for pioneering nitrogen-hungry plants to outcompete *L. maculatus*. Complex plant communities often stem from stressful environments. Paradoxically, nutrient-rich conditions can reduce plant diversity, favouring a few species rather than many.

A successful conservation strategy for a critically endangered species requires multiple elements. A reserve capable of sustaining a viable population is a starting point, focusing physical interventions where they have highest impact. Ex-situ conservation (away from the species' indigenous location) is equally important, forming a contingency in a more stable location.

Viera y Clavijo is the Canary Islands' botanic garden and a valuable ex-situ conservation resource. Home to a seed bank, Viera y Clavijo is a repository for the islands' threatened flora. If prepared correctly, some seed types can be kept in a frozen state for decades, an essential insurance policy against total extinction and a temporal buffer for in-situ conservation to stabilize threatened habitats.

The reality is not always so straightforward. Restoration of threatened species is challenging due to the complex, interdependent nature of plant communities. Simply returning a lost species doesn't always work: the "baseline" of the habitat may have shifted since the plant last occupied it. Previous reintroduction efforts for *Lotus* species on the Canary Islands have failed, but increasingly sophisticated mapping of a species' full potential distribution may highlight new areas for establishment.

The Canary Islands are a microcosm for biodiversity issues: high numbers of endemic species, a considerable array of threats, well-documented flora and fauna and valuable but stretched conservation resources. *L. maculatus* and its fellow Canary Island endemics are wonderful rare plants, quite distinct from any other flora, and more than worthy of conservation.

RIGHT: *Lotus peliorhynchus* (as *Lotus berthelotii*) by Matilda Smith from *Curtis's Botanical Magazine*, 1801.

CRESTED COW-WHEAT

Melampyrum cristatum

Crested cow-wheat is an elegant example of form following function, a plant shaped by a relationship with ants. Highly specific co-evolution brings advantages – the chance to optimize your offer (nectar, food, shelter) for your preferred partners – as well as challenges such as the fragility that comes with dependence on another species. Crested cow-wheat reflects this fragility; it is endangered in Great Britain with a small fragmented population found in just four sites in south-east England.

Crested cow-wheat is a plant of species-rich grassland, one of the UK's most degraded habitats, although its association with the common oak (*Quercus robur*) suggests its habitat was originally woodland glades. It's also found on hedgerow verges, a good analogue of the woodland edge.

Those co-evolutionary forces shaped an exotic, elaborate flower that rewards closer examination. Vivid purple and yellow-tipped tubular flowers (a collective structure called the corolla) emerge from sharply serrated bracts reminiscent of technicolour Venus flytraps. Leaves are long, paired and slender.

Despite the arresting display of the flower, this plant's seeds are its real talking point. Crested cow-wheat evolved to harness the willing industry of the ant through a combination of incentive and deception. Seeds resemble ant cocoons, stimulating their insect collaborator to remove them from the plant's vicinity to store in their underground nests. To give the seeds extra appeal, one end features a nutritious, oily structure called an elaiosome, which is highly palatable to the ant.

OPPOSITE: *Melampyrum cristatum* from Albert Dietrich, *Flora regni borussici*, 1833–44.

RIGHT: *Melampyrum cristatum* from James Sowerby, *English Botany*, 1869.

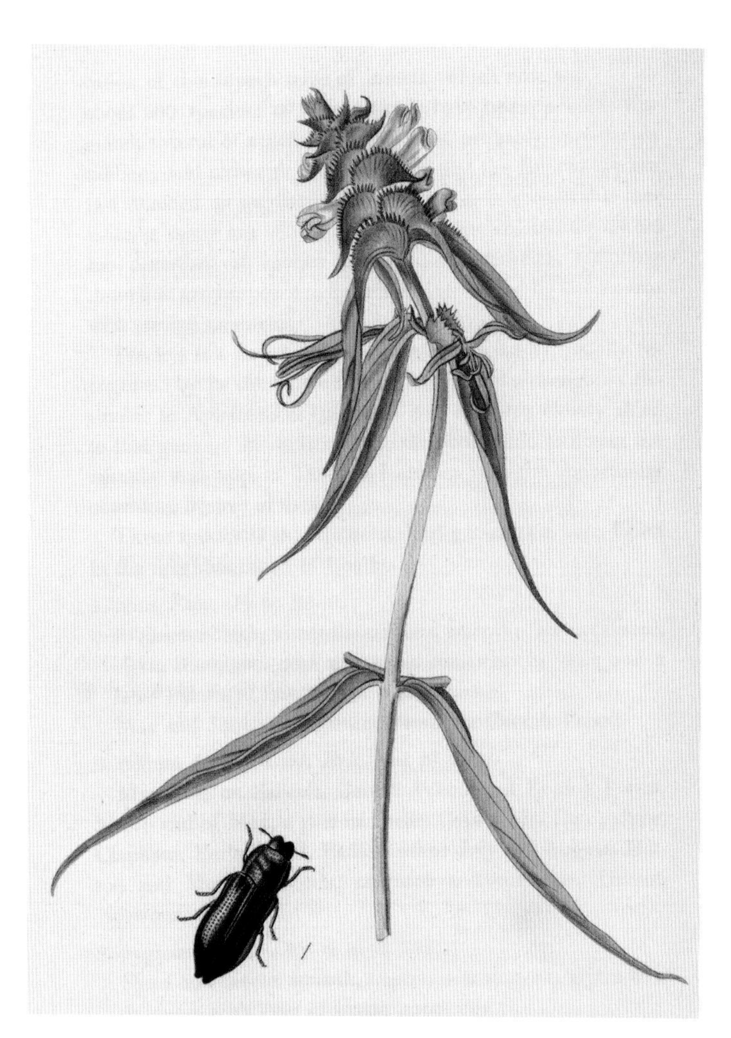

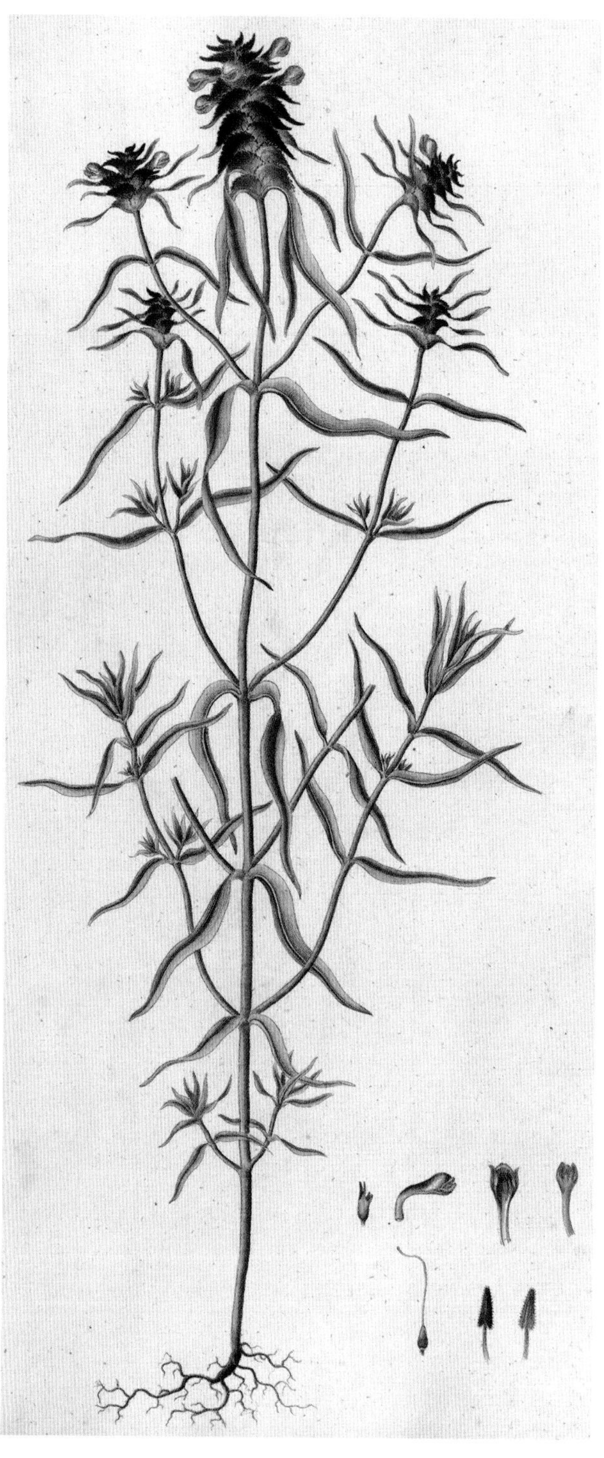

Distributing seeds far from the parent plant is an advantageous strategy, minimizing competition for resources between parent and offspring and, in theory, supporting an ever-growing distribution. So why has this species become so marginalized? Crested cow-wheat is not only dependent on ants to thrive. It's a hemi-parasite (a half-parasite) dependent on host plants for specific resources, especially carbon. This additional layer of dependence increases the vulnerability of *Melampyrum cristatum* to change – if its host plant (typically meadow grasses) reduces, a lack of alternative hosts causes population decline.

With only four populations left in the UK, crested cow-wheat needs urgent conservation. Their species-rich grassland home provides a rich array of ecosystem services to those who live around them. Carbon sequestration, pollinators and uplifting beauty are highly desirable goods, and public funding should support the land management creating them. Maintaining hedgerow verges as diverse habitats, and not spraying them to become monocultures, brings benefits to people and nature alike.

ABOVE LEFT: *Melampyrum cristatum* from John Curtis, *British Entomology*, 1823–40.

ABOVE: *Melampyrum cristatum* from Georg Christian Oeder *et al., Flora Danica*, 1761–1883.

OPPOSITE: *Melampyrum cristatum* from Carl Axel Magnus Lindman, *Bilder ur Nordens Flora*, 1922–6.

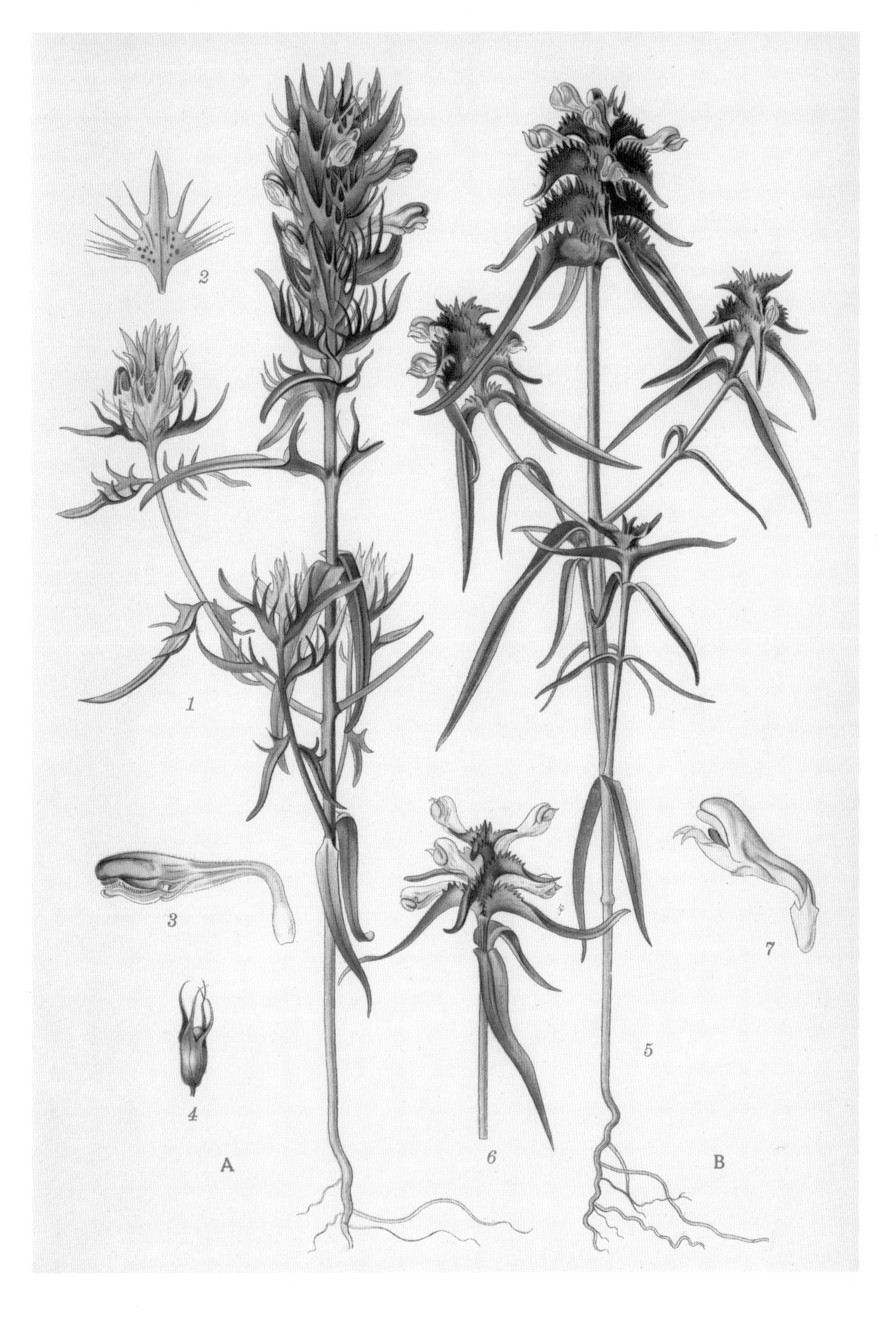
2
1
3
7
4
5
A
6
B

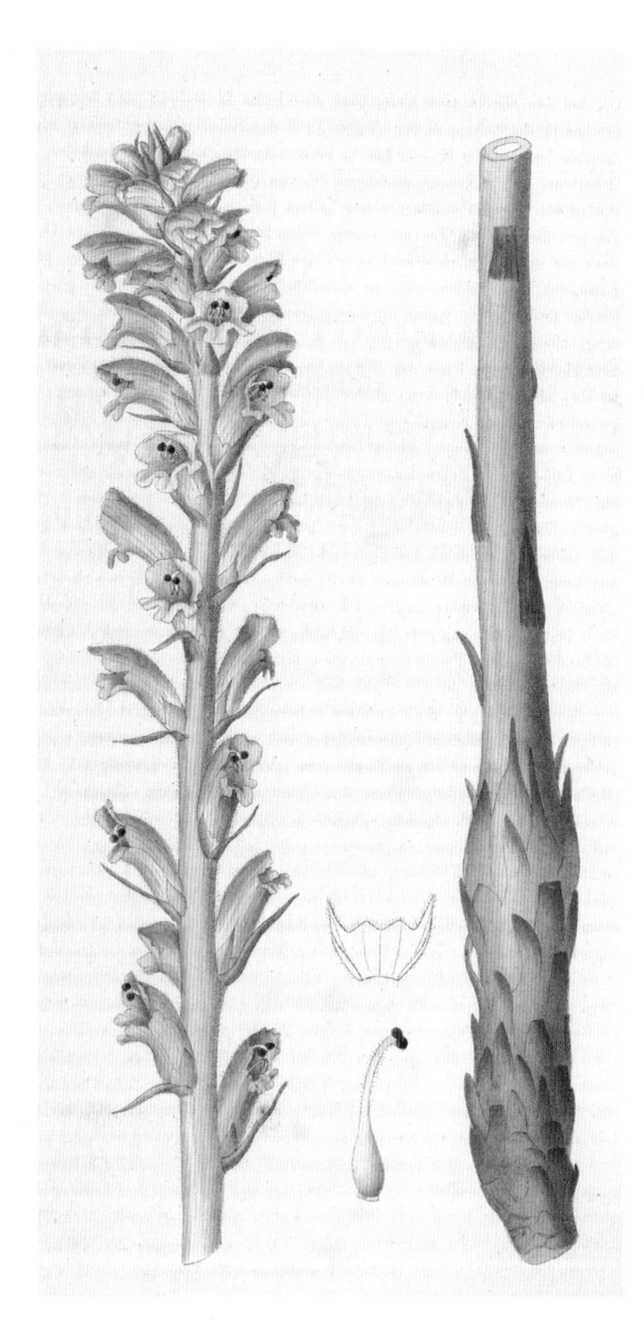

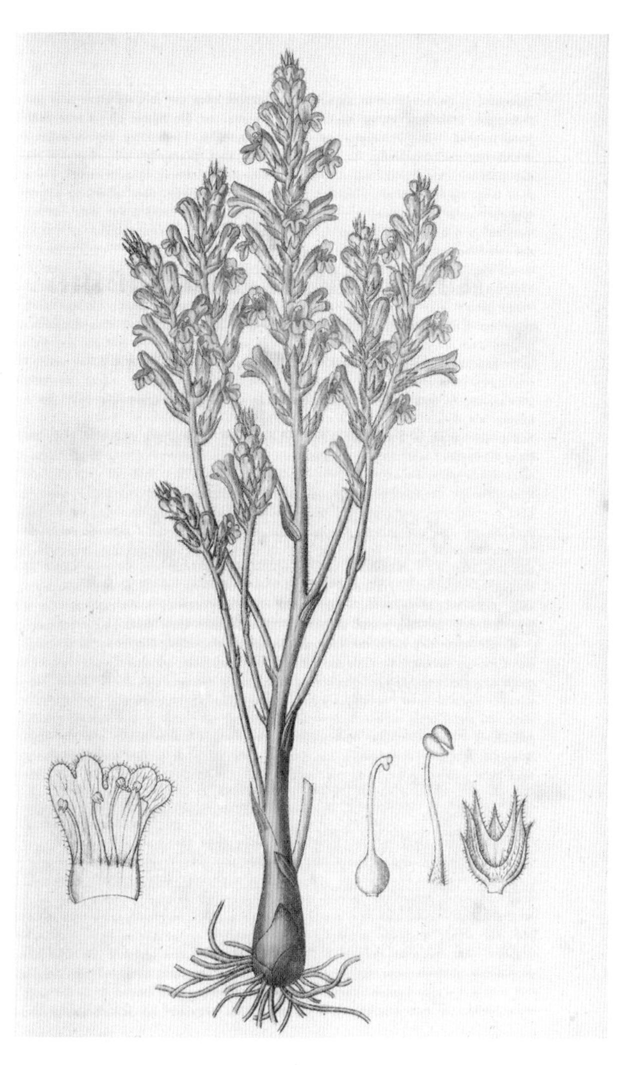

LEFT: *Orobanche caryophyllacea* (as *Orobanche torquata*) from Albert Dietrich, *Flora regni borussici*, 1833–44.

ABOVE: *Orobanche ramosa* from Albert Dietrich, *Flora regni borussici*, 1836.

Crested cow-wheat is in the Orobanchaceae family, a most polarizing group of plants. Across this diverse and curious range of parasites and hemi-parasites are examples of extreme fragility and extreme weediness. Species such as *Orobanche aegyptiaca*, *Orobanche ramosa* and *Orobanche cumana* utilize parasitic prowess to critically weaken crops, especially tobacco, tomatoes, lentils and alfalfa. Their economic impact is considerable in countries heavily reliant on agriculture, with crop losses of up to 75 per cent recorded.

This extraordinary spectrum of ecological adaptability shows there's no greater evolutionary force than human impact. The Anthropocene has been described as the era of human devastation – the first period of mass extinction caused by a single species. There are too many examples of wonderful species lost to our lack of foresight, but there's nuance to this story. Our interventions have also favoured species too, with Orobanche species shifting from dependent, stable niche dwellers to rampant weeds because of an unanticipated ability to adapt to our disruptive actions.

To understand this ecological shift is to understand the relationship between biodiversity and our own economic prosperity, in which our best interventions can benefit both humans and nature.

D. Auguſti Qvirini Rivini
Lipſienſis,
INTRODUCTIO
Generalis
in
REM HERBARIAM.

M DC XC.
Sumptibus Autoris,
Lipſiæ,
Typis CHRISTOPH. GÜNTHERI.

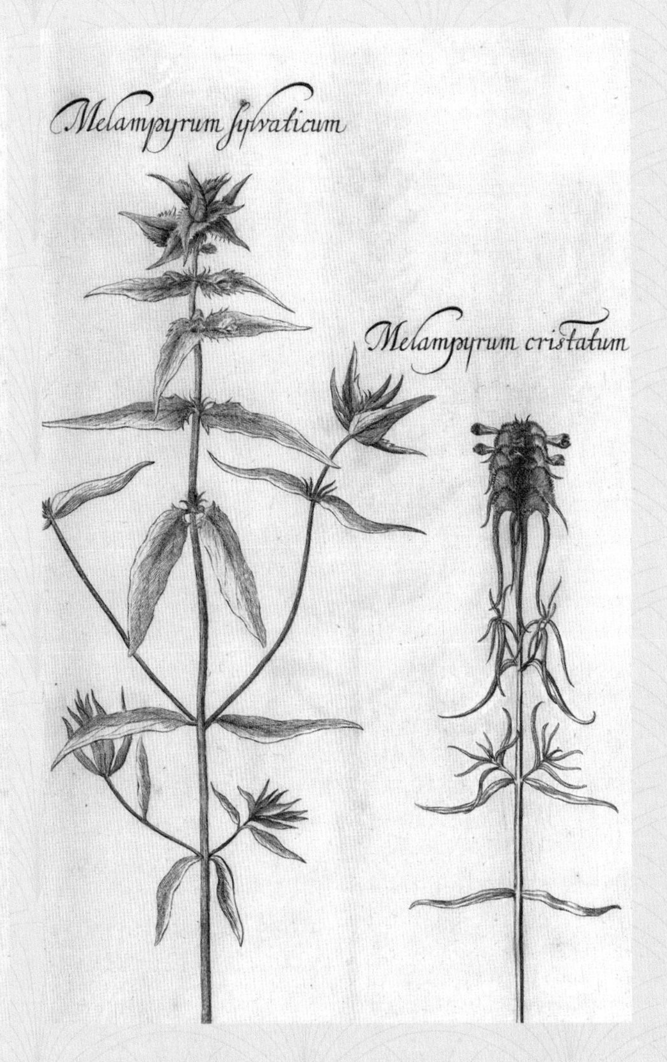

Title page and page featuring *Melampyrum cristatum* from August Quirinus Rivinus, *Introductio generalis in rem herbariam*, 1690–9.

Rivinius was a German botanist whose work on plant classification was later used by Linnaeus to help form his binomial taxonomic classification system.

27

ATTENBOROUGH'S PITCHER PLANT

Nepenthes attenboroughii

The pitcher plant (several species of *Nepenthes*) is a precisely evolved, curiously beautiful instrument of death. Capable of ensnaring anything from an insect to a mammal (rumours persist of a rat-catching pitcher plant), it is a plant redolent of its highly complex, highly interactive tropical forest habitat.

Nepenthes species are often lumped, rather clumsily, into a group known as the carnivorous plants. It's a diverse range from across the evolutionary spectrum, but all are dependent, to varying degrees, on supplementary nutrients from the decomposing fauna they catch. From the overt drama of the Venus flytrap to the microscopic, subterranean peril of the bladder-trapped *Utricularia* genus, they present a fascinating set of adaptations to the singular issue of needing more nutrition.

Nepenthes is a substantial genus composed of over 170 species. Its distribution is across the highly biodiverse old-world tropics, with notable concentrations in the Philippines, Borneo and Sumatra. The inaccessibility and complexity of these habitats means new species of pitcher plants are regularly being discovered, with 12 new finds named in 2013 in the Philippines alone.

These plants are predominantly climbers (although there are several shrubby species), stems of several metres scrambling up the vertiginous trunks of rainforest trees. They're dependent on the structural support of their fellow forest plants, but don't parasitize them, deploying a masterpiece of evolution to gain their supplementary nutrients.

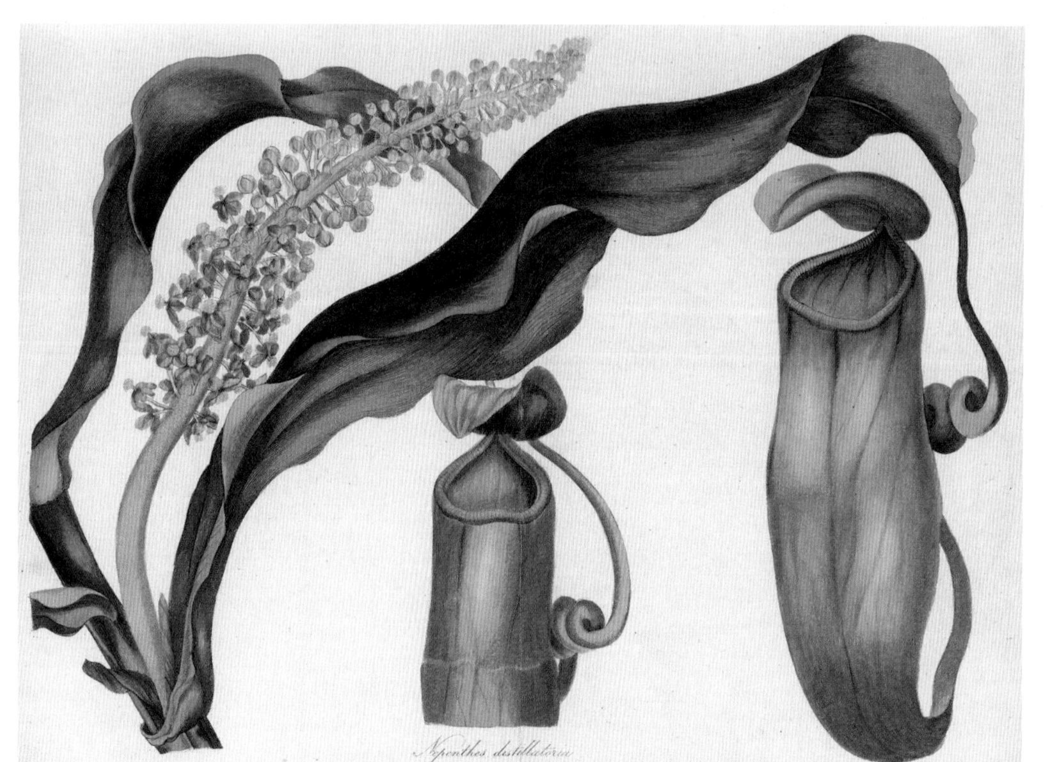

LEFT: *Nepenthes distillatoria* by S. Holden from Joseph Paxton, *Paxton's Magazine of Botany and Register of Flowering Plants*, 1841.

OPPOSITE: *Nepenthes attenboroughii* by Lucy Smith, 2016. © The Carnivorous Plant Society.

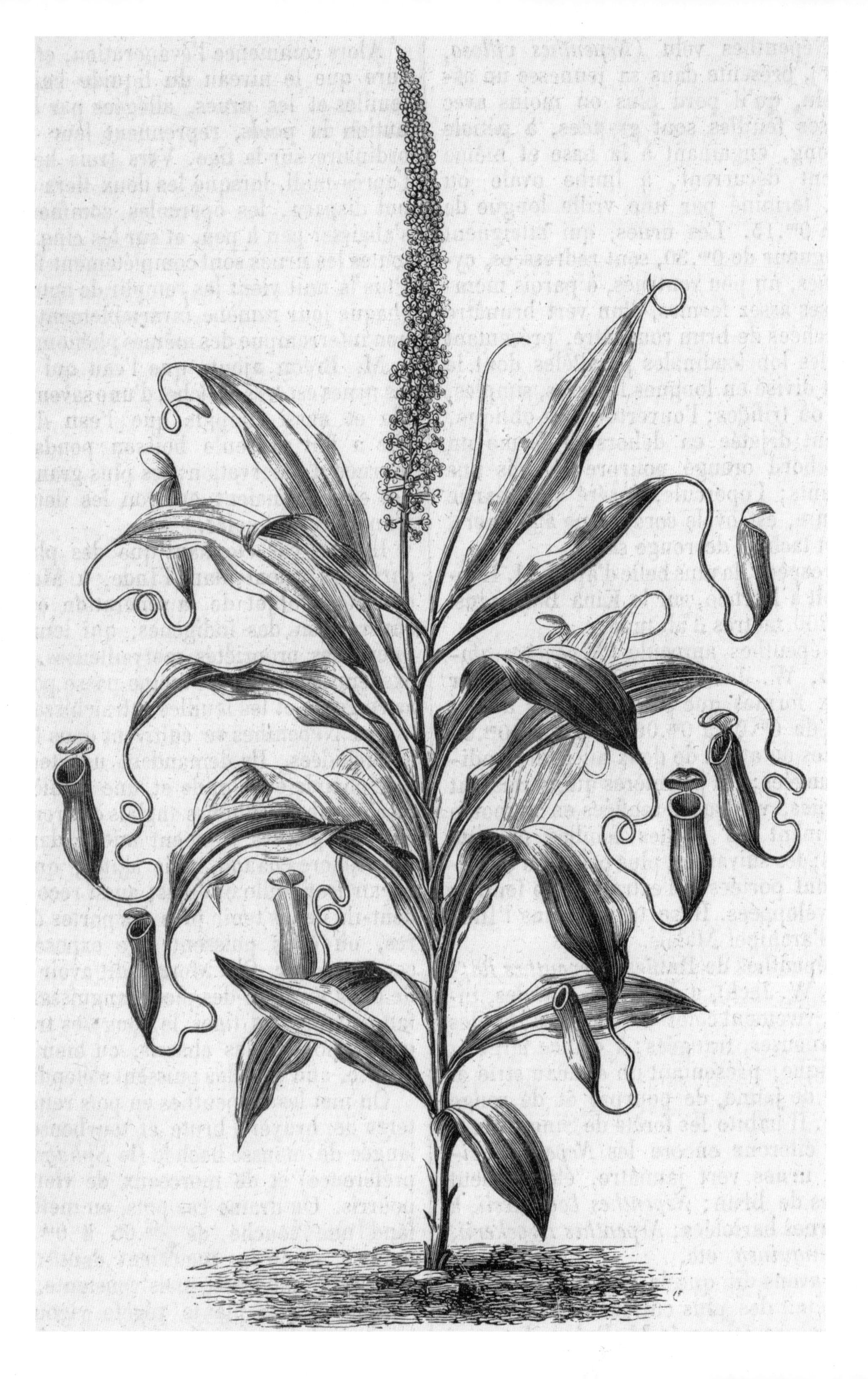

An extraordinary set of adaptations lure, trap and kill *Nepenthes*'s prey. The pitchers deploy a range of phytochemical signals to attract insects and mammals, emphasizing the sweet reward lurking within the nectaries inside the pitcher. The inner rim of the pitcher is exceptionally smooth, capable of inducing prey into an uncontrolled tumble. Descending into the pitcher is far easier than getting out: an array of downward-facing hairs inhibit climbing, with repeated attempts slowly tiring the incumbent. When they finally fall to the bottom, a pool of enzyme-rich liquid is ready to start the process of digestion, reducing the prey to its constituent elements.

Nepenthes attenboroughii, named in honour of naturalist Sir David Attenborough, is a botanical newcomer and a reminder that conserving the world's biodiversity, including naming new species, is an unrelenting challenge. The Philippines are a hotspot for *Nepenthes* endemism, with 16 species occurring only in the archipelago. The extraordinary remoteness and extent of their forest habitat stirs a tantalizing sense that there's more to be discovered.

Indeed, this was the motivation for a 2007 expedition to Mount Victoria, in the Palawan archipelago. The expedition started in dense forest that thinned as the team climbed, suddenly opening to a rocky habitat with scrub vegetation. Growing on rocky protrusions amid the scrub was something unusual: a *Nepenthes* of dramatic dimensions, with pitchers up to 30 centimetres long. Newspapers seized on the impact of a new plant discovery, but headlines were frenzied: "New meat-eating plant discovered", "Rat-eating plant found on mountain".

The botanists' instincts that this species was unknown to science were quickly confirmed. A new species, and a new entry to the IUCN Red List: *N. attenboroughii* was immediately classified as "Critically Endangered". Covering a mere 10 square kilometres on one mountain, on one island in the Philippine archipelago, the population of *N. attenboroughii* already hangs by a thread.

Its newfound fame is not universally welcomed. A desirable new species is now on the market for unscrupulous collectors and poaching is a serious threat for this tiny population. How many more remote mountaintops harbour species alien to science, plants that could transform our understanding of evolution or provide extraordinary benefits? The challenge to identify and conserve continues apace.

OPPOSITE: *Nepenthes distillatoria* from *Revue horticole*, 1861.

RIGHT: *Ceylon Pitcher Plant and Butterflies*, by Marianne North, 1877.

University of Pennsylvania.

THE COLLEGE.

BOTANY.

PHILADELPHIA, 31st August. 1907.

Dear Dr Prain

I have just returned from the Catskills where for some weeks I have been bagging innocent game in the form of dried plants. It is now my pleasant duty to thank you for the Nepenthes plants, you and Mr Watson kindly sent in my absence. All are doing well, and are a welcome addition to our collection, as I am hoping soon to go rather minutely into Nepenthes hybrids.

My Sarraceniaceae and Nepenthaceae monographs for Engler are now almost ready for publication, and I am now working on a more exhaustive monograph of the whole series as I had planned years ago. In

furtherance of this I am hoping to make another trip to Kew, Leiden and other centres rich in material, next summer if all goes well.

I hope you do not feel very bad against me that I have tempted whole three of your Kewites over here to our Botanic Garden. It is practically impossible to get skilled native help here, for the average Yankee thinks "there aint enough dollars in that ere business for me". Thomas seems to be a very nice intelligent fellow. I was pleased with him last year when I happened to meet him in his rooms, and to get some help in plant preservation from him. The other two also shape well.

Though full of years and honors I was very sorry to note the death of Dr Masters, one who always seemed to me a noble kindly gentleman. I am

University of Pennsylvania. 399

THE COLLEGE.

BOTANY.

PHILADELPHIA,

glad however to note that Sir Joseph has passed the 90 milestone in life's journey, a long life of truly marvellous activity.

Give my kindest remembrances please to your good Lady, and the different workers at Kew whose acquaintance or friendship I have had the good fortune to make. Best wishes also for your highest success in your onerous but most honorable position.

Very truly Yours

John M. Macfarlane

Letter from John Muirhead Macfarlane, University of Pennsylvania, Department of Botany, Philadelphia, USA, to Sir David Prain, Director of Kew, 31 August 1907.

Macfarlane was a Scottish botanist who spent nearly 30 years as a professor at the University of Pennsylvania. In 1908 he published a complete account of the *Nepenthes*, the most comprehensive account of the genus at the time. He mentions this work in his letter to Prain and reports that it is nearly complete.

28

LEAST WATER LILY

Nuphar pumila

The least water lily is well named, a plant that can never be accused of ubiquity. Confined in England to a single body of water, a mere in the Midlands county of Shropshire, it's a neat prime example of the niche specialization some species have evolved.

Consider the omnipresence of plants like *Buddleja* (butterfly bush) or dandelions. They're capable of growing, flowering and setting seed in dry or damp soil – regardless of how rich or poor in nutrients that soil is – and are tolerant of both sun and shade. These plants have a broad ecological bandwidth, allowing them to become pioneering in nature and cosmopolitan in distribution, a mode of growth some may describe as "weedy".

Human activity is often noted as a pressure on biodiversity throughout this book, but it can favour some plants. Disturbed ground, a constant by-product of humanity's development, offers opportunities to pioneering species. It stimulates a proliferation of seeding or vegetative colonization (spreading through roots or overground stems), often spotted by railway embankments and riverbanks.

Nuphar pumila will never dominate the landscape like the dandelion. It needs precise levels of light, oxygen, nutrients and turbidity (the degree to which its watery habitat moves) and is incapable of growing on land. Fundamentally restricted to a set of infrequently occurring conditions, the least water lily is a fragile species, susceptible to environmental perturbation.

While not as conspicuously decorative as the cultivated forms that were so beloved by Monet, *N. pumila* is instantly recognizable as a water lily, with distinctive wavy-edged, heart-shaped floating leaves that rise from a submersed rhizome (perennial root structure). Flowers are subtle but beautiful: a simple structure reflective of the plant's ancient origins, in pale yellow.

A key characteristic of this plant is the hairy, or pubescent, underside to the leaf, a trait that distinguishes it from the closely related *Nuphar lutea*. Established wild populations are a dramatic sight, massed near the edges of large freshwater bodies, their leaves beating a languid rhythm as breezes zip across the water.

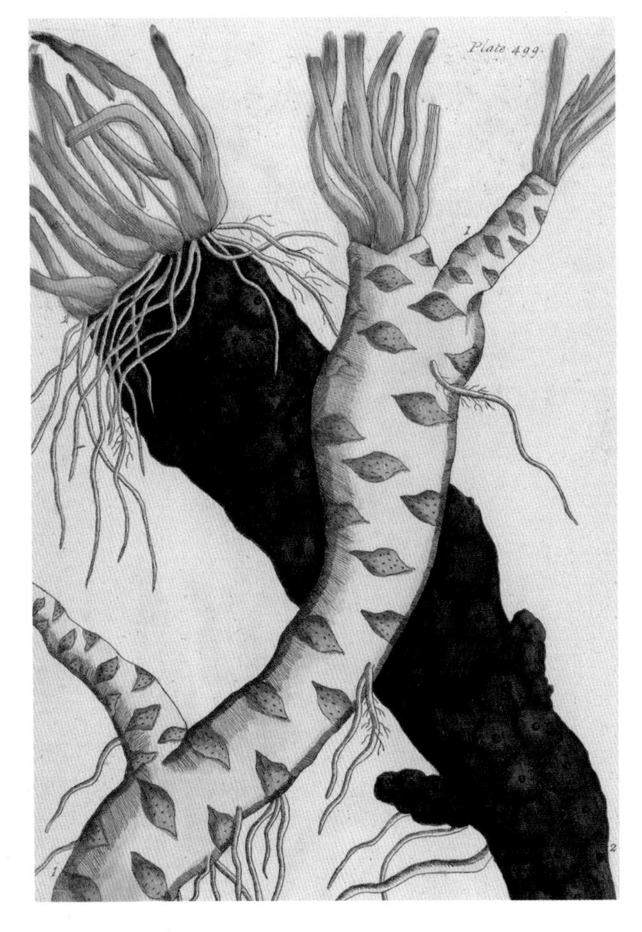

OPPOSITE: *Nuphar pumila* from James Sowerby, *English Botany*, 1863.

ABOVE: Roots of *Nuphar lutea* from Elizabeth Blackwell, *A Curious Herbal*, 1737.

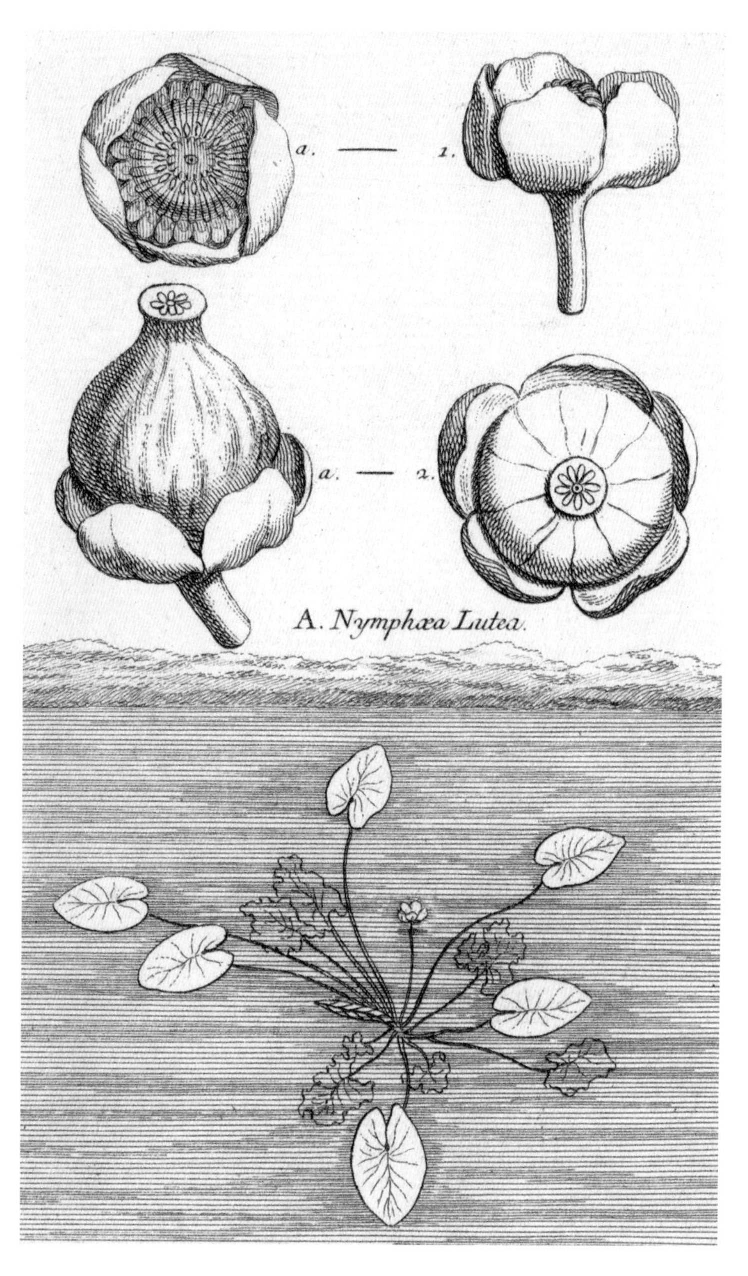

ABOVE: *Nuphar lutea* (as *Nymphaea lutea*) from Etienne-François Geoffroy and François Alexandre Pierre de Garsault, *Description, vertus et usages de sept ceuts dix-neuf plantes, tant étrangeères que de nos climats*, 1767.

OPPOSITE: *Nuphar lutea* from Elizabeth Blackwell, *Herbarium Blackwellianum*, 1760.

Prior to a conservation programme initiated in 2017, the least water lily's only English distribution was in the waters of Colemere, near Whitchurch in Shropshire, close to the Welsh border. The isolation of this population – a sole fragment hundreds of miles from more extensive distributions (over a hundred recorded populations) in Scotland – poses interesting questions for scientists. Were the Scottish and English populations once part of a continuous whole? Did an extraordinary force, such as a rapidly retreating glacier, leave the English population stranded?

Geneticists can sample and compare the genomes of populations to look for indications of divergence and drift. Big genetic differences may indicate that the populations separated a long time ago, although environmental influences or hybridization with similar species *N. lutea* muddy the genetic waters and hamper the quest for a definitive conclusion. Initial analysis indicates that the English and Scottish populations have similar genomes.

The least water lily is threatened in several countries across its European distribution. It is intolerant of eutrophication (an excessive richness of nutrients), and intensive agriculture upland of the water lily's habitat causes nutrients to wash out of the land and into the water. Connected land management policies that consider the environmental interactions across entire catchments are a step towards protecting species with fragile ecologies like *N. pumila*.

While the least water lily ended up in a sole English site, records indicate it once held additional strongholds in England. In a study sponsored by Natural England, researchers proposed other sites that have the environmental conditions capable of nurturing the least water lily. A conservation programme was launched to increase the plant's English distribution and, perhaps, give rise to a new common name.

A team including scientists and horticulturists from the Royal Botanic Gardens, Kew, braved the chilly, rather murky waters of Colemere to collect seeds and root material to propagate and develop into an ex-situ (away from the initial wild site) collection for introduction to further water bodies.

How do you grow a plant with the precise environmental requirements of the least water lily? The ingenious horticulturists at Wakehurst, Kew's wild botanic garden, used their aquatic plant know-how and first-hand experience of plumbing Colemere's depths to simulate life under Shropshire's waters. Tanks with moderate light, minimal nutrition, just enough oxygen and a steady but cool temperature allowed this fragile species to flourish, creating the plants that will form new populations of this beautiful, rather particular plant.

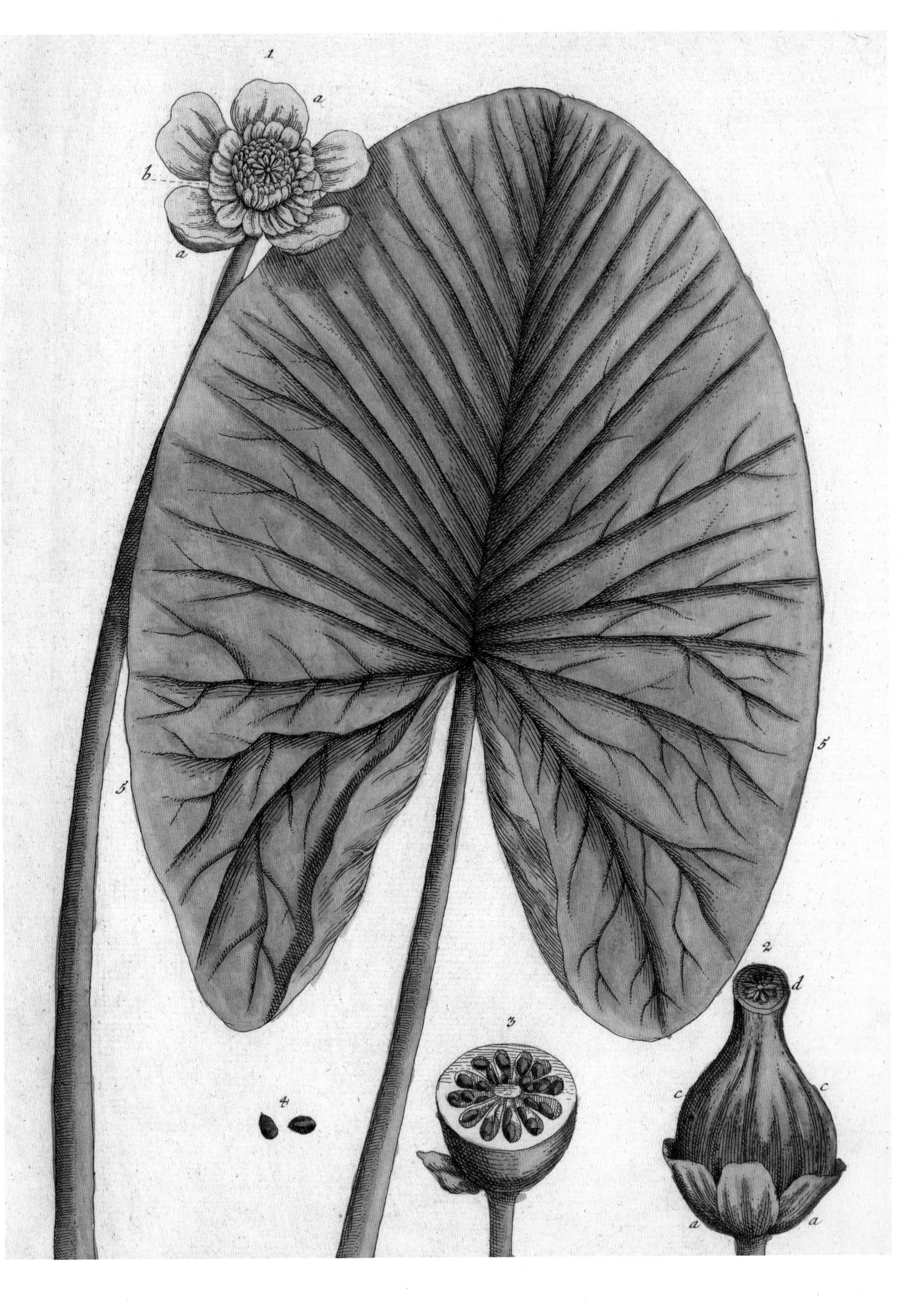
1
a
b
a
5
5
2
d
3
c
c
4
a
a

29

THERMAL WATER LILY

Nymphaea thermarum

Water lilies are an ancient group, one of the earliest flowering plants to evolve. The long-standing origins of this family are revealed by their distribution throughout Africa, Asia, South America and Europe, an indication the family started to branch when some of our current continents were still conjoined.

Far from the least water lily's chilly Shropshire mere lived another extraordinary member of the family, with equally specific requirements. The thermal water lily (*Nymphaea thermarum*) was found in just one wild location: a thermal hot spring in Mashyuza, south-west Rwanda. Specimens didn't grow in the hot spring itself, but in the warm, wet mud moistened by its outflow. Preferring a steady temperature of 25°C, this plant defines the term "niche requirements".

N. thermarum was only discovered in 1987 and was a notable feat for the botanist responsible, given its singular location and diminutive size. The smallest of the water lilies, its tiny lily pads measure as little as a centimetre across with white flowers, centred with a cluster of yellow stamens borne a few centimetres above the leaf surface.

A plant in need of such precision is vulnerable to change and, perhaps inevitably, shifting land use surrounding Mashyuza imperilled *N. thermarum*. Overuse of the aquifer changed the hydrology of the water lily's habitat, drying its nurturing mud and starting a terminal decline that took less than three years. The thermal water lily is now extinct in the wild, a sobering reflection on the fragility of endemic plants and a rousing challenge for global plant conservation. Thankfully, the plant is not lost to us entirely – the ingenuity of horticulturists keeps the thermal water lily alive in our botanic gardens.

Replicating mud consistently moistened by a thermal spring is not everyday work for horticulturists. Luckily, the Royal Botanic Gardens, Kew, is no ordinary garden. The Tropical Nursery is one of the wonders of the botanic garden world, tucked away from visitors and home to an extraordinary collection of rare plants, housed in bespoke growing conditions. The team in the Tropical Nursery face a regular challenge: how do you grow something that's never been grown before? Finding the answer could save a plant from extinction.

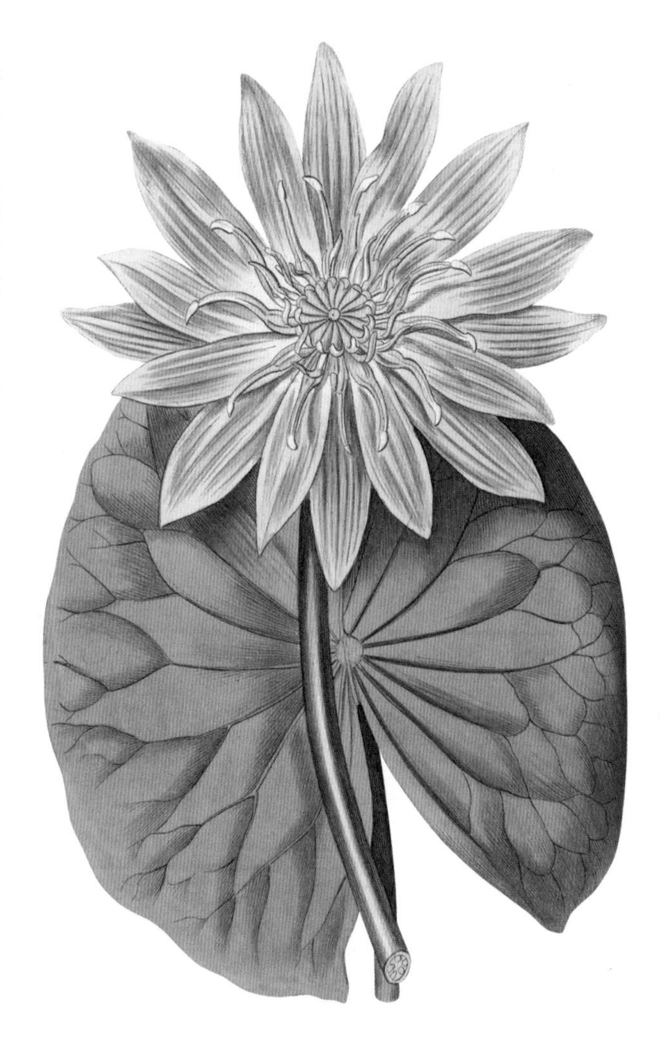

ABOVE: *Nymphaea nouchali* (as *Nymphaea stellata*) from *Curtis's Botanical Magazine*, 1801.

OPPOSITE: *Nymphaea thermarum* by Lucy Smith from *Curtis's Botanical Magazine*, 1984.

Deducing how to grow *N. thermarum* fell to Kew Botanical Horticulturist Carlos Magdalena, a master at replicating precise conditions to make rare plants thrive. The plant's wild habitat is the starting point, with every detail considered. Light levels, soil type, pollinators, nutrient availability, ventilation levels and even the mineral content of the water are all wild factors that could nurture or kill the plant in cultivation. Magdalena's aspiration was for *N. thermarum* not just to survive but also to thrive, producing flowers and seeds to create new generations.

Initial experiments grew the precious water lilies in mud, with the temperature carefully maintained at a steady 25°C. The pH and dissolved gas levels were carefully monitored and, despite the precise environmental manipulation, young plants struggled. The breakthrough came when cultivation shifted to growing in cubes of compost, surrounded by water. While not an exact match for wild conditions, it allowed more precise regulation of the levels of carbon dioxide and oxygen. With this switch, the Tropical Nursery soon had a flourishing population of thermal water lilies, and they flowered for the first time in 2009. With a reliable propagation method and global network of collaborative botanic gardens available, it's possible to build a viable new population of *N. thermarum*.

Restoration to the wild, that isolated thermal spring in south-west Rwanda, is possible, but only in partnership with local policymakers and land managers. Until that moment comes, it remains safe with Kew.

ABOVE: *Nymphaea alba* by Matilda Smith from *Curtis's Botanical Magazine*, 1926. The coloured and uncoloured plates are shown here.

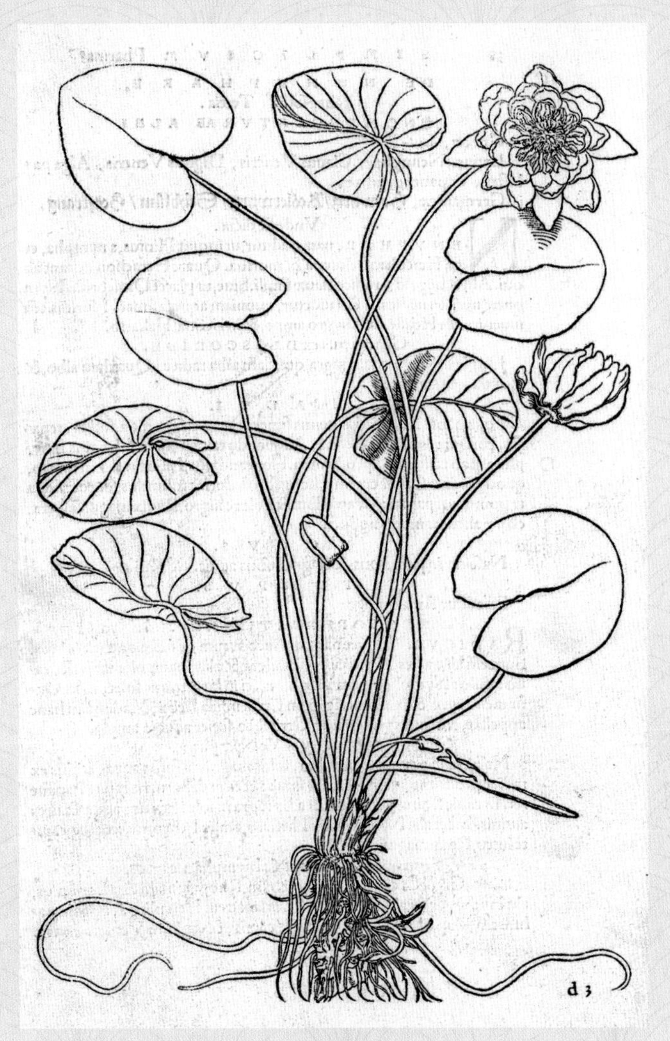

LEFT: *Nymphaea alba* (as *Nenuphar alba*) from Otto Brunfels, *Herbarium tomis tribus*, 1530. Wellcome Collection.

BELOW: *Nymphaea thermarum* by Gustavo Surlo from *Curtis's Botanical Magazine*, 2019.

PRICKLY PEAR CACTUS

Opuntia

Used for food, medicine, dyes, construction and enclosures, the prickly pear cactus is an immensely helpful resource for humans. Although originally from the Americas, it is globally abundant in the form of *Opuntia ficus-indica*. This most cosmopolitan of prickly pears is assimilated into cuisines and cultures of the Mediterranean, Middle East and North Africa, but other *Opuntia* species are less abundant. *Opuntia galapageia* occurs only on the Galápagos Islands, a population stabilized by recent conservation efforts, while *Opuntia chaffeyi* is limited to one dried-up Mexican lake.

Opuntia is a bounteous natural resource: easy to cultivate with minimal water requirements and diverse in its economic applications. The stout, bristling form of *O. ficus-indica* makes an excellent barricade to cattle and was used to demarcate field boundaries in Malta. *O. ficus-indica*'s willingness to adapt readily to new environments tips it into the role of invader – albeit one with desirable properties many are keen to harness.

The economically motivated introduction of *Opuntia stricta* (erect prickly pear) into Australia quickly backfired. Rather than benefiting from its intended purpose as a stock barricade and host plant for cochineal beetles, vast tracts of potential farmland (over 40,000 square kilometres) were quickly choked. Its subsequent eradication – by introduction of an *Opuntia*-feeding moth that quickly arrested the plant's dominance – was a rare triumph for biological control in reversing a rampant alien species.

Once the spiky hostility of the fruits is penetrated (the finer spines, known as glochids, can lodge painfully in the throat), the nutritious flesh of *O. ficus-indica* is a good source of vitamins C, E, and K, antioxidants and amino acids. Prickly pear is a staple of Mexican cuisine, with a wide range of sweet and savoury applications.

OPPOSITE: *Opuntia decumbens* by Walter Hood Fitch from *Curtis's Botanical Magazine*, 1842.

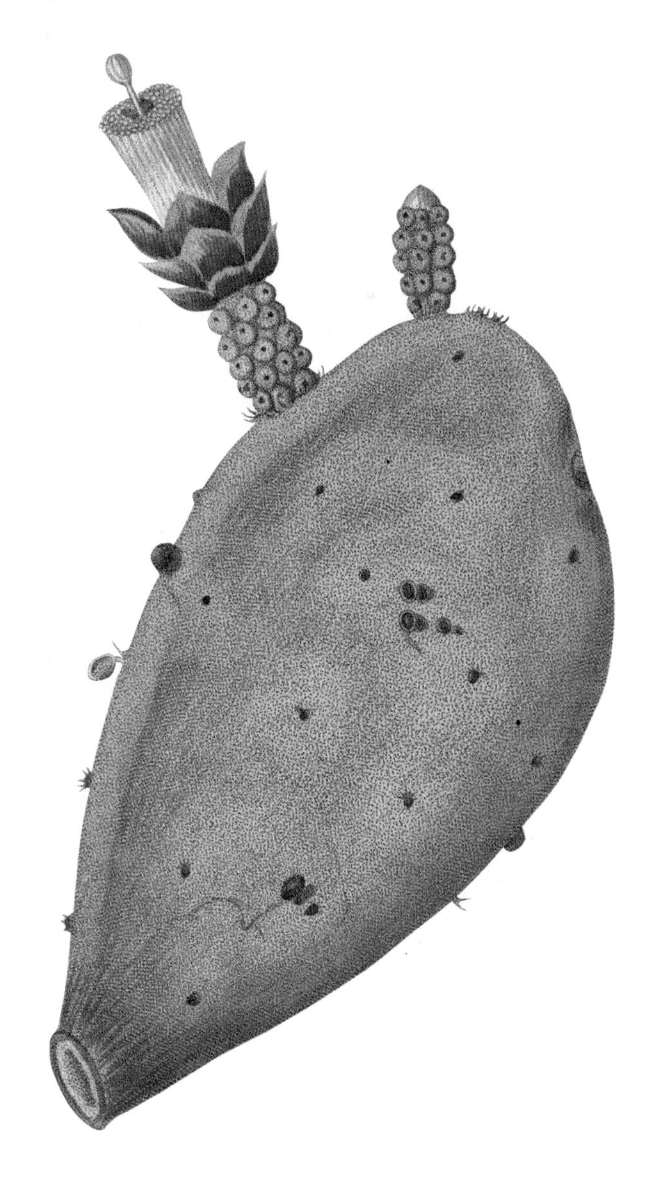

RIGHT: *Opuntia cochenillifera* (as *Cactus cochenillifera*) from Michel Étienne Descourtilz, *Flore médicale des Antilles*, 1821–9.

Opuntia is well established in traditional Mexican medicine for the treatment of burns, obesity, diabetes and gastric ailments. A recently synthesized product based on the cactus is treating metabolism disorders, and there's growing interest in its ability to lower cholesterol, limit atherogenesis (the build-up of fatty deposits) and improve heart health. The prickly pear's drought tolerance and medicinal potential make it a future crop of great interest.

The contrast between *O. ficus-indica*'s abundance and *O. chaffeyi*'s scarcity couldn't be starker. The latter is a small cactus of delicate form, with slender branching stems growing no more than 25 centimetres in length. Growing in the arid Mexican state of Zacatecas, it appears well adapted to its harsh environment, brandishing long spines evolved to minimize evaporation and deter grazing. Its distribution is highly localized: *O. chaffeyi* only occupies a tract of 12 square kilometres in the wild, primarily the flood plain of a dry lake, and there are only 15 known mature specimens.

Human activity is rapidly diminishing a species highly dependent on a settled environment. Modern cattle breeds graze less discriminately than the animals *O. chaffeyi* evolved with, so its spiny defences are readily breached. The thick fleshy roots are used in traditional medicine for their anti-inflammatory properties, leading to destructive harvesting and illegal trading, in contradiction of *O. chaffeyi*'s CITES (Convention on International Trade in Endangered Species) listing.

Opuntia galapageia is found only on the Galápagos Islands. An extraordinary arborescent (tree-like) cactus growing up to 5 metres with a distinct "trunk", it would tower over diminutive *O. chaffeyi*, and is eaten by the islands' tortoises. Although introduced grazers were threatening the viability of the population, proactive conservation efforts have now arrested this threat.

The Galápagos Islands show how biodiversity can be correctly valued to support conservation, with those lucky enough to visit paying a suitable financial premium. The income generated supports Galápagos conservation work, maintaining one of the world's wildlife wonders through a virtuous feedback loop.

RIGHT: *Opuntia* by J. Pass, *c.* 1800. Wellcome Collection.

OPPOSITE: *Aloe and Cochineal Cactus in Flower, Teneriffe*, by Marianne North, 1875.

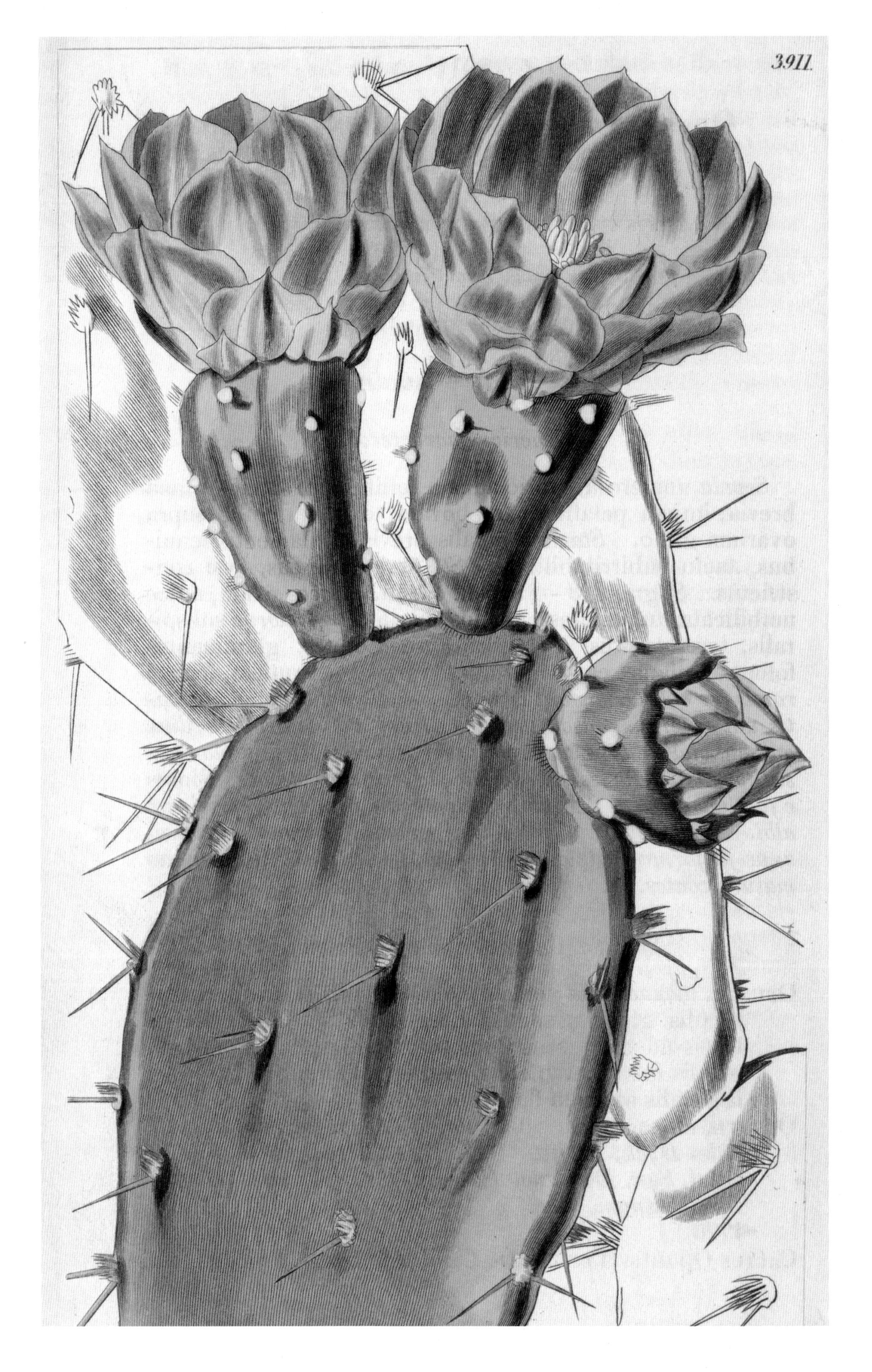
3911.

REMBERTI
DODONAEI
MECHLINIENSIS
MEDICI CÆSAREI

STIRPIVM HISTORIAE
PEMPTADES SEX.
SIVE
LIBRI XXX.

ANTVERPIÆ,
Ex officina Christophori Plantini.
M. D. LXXXIII.

PEMPTADIS SEXTÆ LIB. III. 801

Siccæ autem Ficus, siue Caricæ, mediocriter excalfaciunt, nutriunt quoque, & alimenti non exiguum conferunt, sed tamen non probum sanguinem gignunt: quamobrem & offenduntur qui iis largius vescuntur, & copia pediculorum ipsis prouenit: Aluum verò & faciliorem reddunt, & ad excretionem irritant; præsertim ante alios cibos præsumptæ. Asperæ arteriæ, pectori, pulmoni, renibus, ac vesicæ vtiles sunt, nec muliebri vtero incommodæ: attenuandi siquidem, leniendi, concoquendi, ac maturandi facultatem obtinent. Quæ pectori inhærent concoquunt, maturant, & expurgant, præsertim cum Hyssopi aut consimilis herbæ decocto: ventri, tussi, ac diuturnis pectoris ac pulmonis vitiis hoc modo auxiliantur. renes expurgant, & ab arenulis liberant: vesicæ dolores mitigant: grauidæ si Ficubus aliquot ante partum diebus vescuntur facilius pariunt: exanthemata siue papulas præterea, & sudores Ficus etiam, vt Dioscorides ait, prouocant. Tumores vel concoquunt vel discutiunt; ad concoquendum Triticea farina; ad digerendum Hordeacea admixta, Galeno auctore, conferunt.

Decoctum Ficuum oris ac faucium tumores & abscessus citò facileq; ad suppurationem perducit, sæpius in ore detentum aut gargarizatum.

Lacteus autem Fici arboris liquor & succus foliorum eius non mordicat tantùm, aut abstergit vehementer, sed & vlcerat, & ora vasorum reserat, & myrmecias eiicit: verùm & purgare potest: auctor Galenus. Coaguli modo, Dios. ait, lac coagulat, concretumque vt acetum dissoluit. Menstrua cum oui luteo aut Tyrrhenica cera appositus cit: podagricorum cataplasmatis cum farina Fœnigræci & aceto vtilis. lepras, lichenas, discolorem à Sole cutem, alphos, scabies, ichoras cum polenta expurgat. Instillatum plagæ percussis à scorpione, venenatorum ictibus, & rabiosi canis morsibus auxiliatur. dentium doloribus lana exceptus prodest, & cauitati eorum inditus. myrmeciam tollit cum adipe circumlita in ambitu carne.

De Ficu Indica. CAP. XXI.

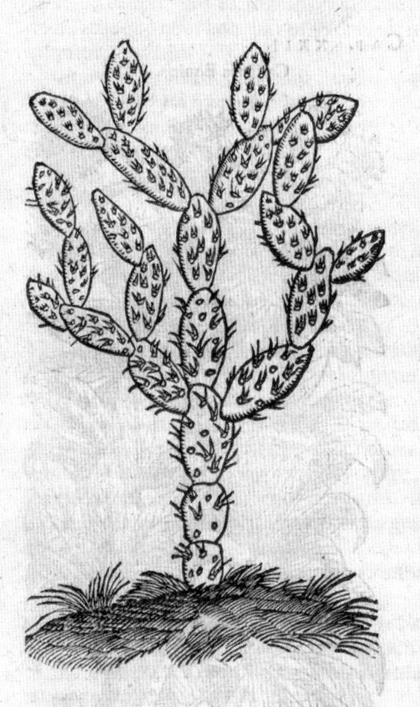

Ficus Indica.

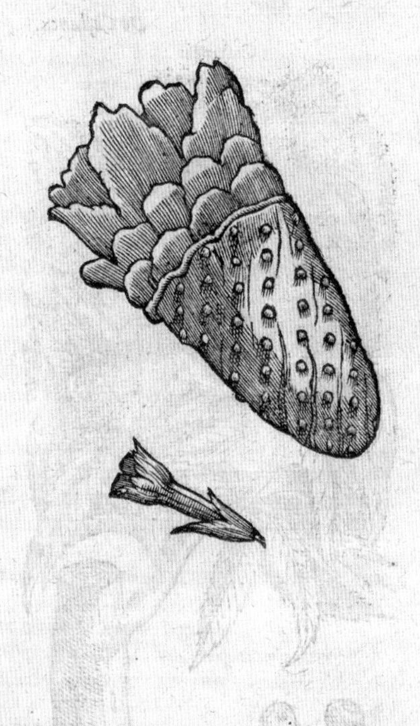

Ficus Indicæ fructus & flos.

INDICA hæc Ficus appellata arboris specie absque caudice & ramis adolescere videtur, tota enim nihil aliud quàm ex foliis folia apparet; folium siquidem in terram depactum mox radices agit, & alia ex se promit, à quibus & aliis prodeuntibus succedunt rursus & alia, donec ad arboris altitudinem peruenerint, effusis interim & veluti ramis, ab vno nempe subinde

Title page and page featuring *Opuntia* (as *Ficus*) from Rembert Dodoens, *Remberti Dodonaei*, or *Stirpium historiae pemptades sex, sive libri XXX*, 1583.

Dodoens' monumental work was an international hit, published in several languages, and at the time, the second most-translated book after the Bible.

OPPOSITE: *Opuntia monacantha* by Walter Hood Fitch from *Curtis's Botanical Magazine*, 1842.

THE EGG-IN-A-NEST ORCHID

Paphiopedilum bellatulum

The orchid family is vast, complex and filled with extraordinary examples of evolutionary sophistication. A total of 31,000 species are found anywhere from English meadows to American bogs to tropical forests, growing terrestrially or aerially. These plants have evolved to drug pollinators with caffeine, stun them with projectile pollen, convince them to mate with flower structures or imprison them until their unwitting role in pollination is complete. They've co-evolved flower structures with a type of moth whose proboscis is 20 centimetres long, they live in symbiosis with a fungus and are capable of existing exclusively underground for years.

Another species whose existence is utterly intertwined with orchids is *Homo sapiens*. We're beholden to the beauty, mystery and usefulness of this diverse family, pushing horticultural boundaries to grow them and collectively spending billions of pounds to acquire them. The trade in orchids focuses on three significant assets: beauty, medicinal properties and flavour.

The climbing orchid *Vanilla planifolia*'s wild origins are in Mexico, where Aztec ruler Montezuma mixed its small aromatic seeds with cacao. It wasn't long before this unassuming orchid's ability to enhance flavour and aroma was recognized beyond its wild home, and vanilla plantations in Madagascar, Réunion and Mauritius constitute some of the world's most valuable agricultural holdings.

Orchids have a long history of medicinal applications, especially in China, Asia Minor and the Middle East. A tea called salep is made from the ground tuber of the genus *Orchis* and is drunk from Turkey to Saudi Arabia to soothe digestive problems. Chinese preparations from the orchid genus *Dendrobium* have been used to alleviate diabetes, while *Bletilla* treats haemorrhage. Research into the efficacy of these traditional uses is ongoing, but compounds isolated from *Bletilla* have proven to be potent coagulants.

Ornamental orchids are a significant global commodity.

LEFT: *Paphiopedilum bellatulum* (as *Cypripedium bellatulum*) by Alphonse Goossens from Alfred Cogniaux, *Dictionnaire iconographique des orchidées*, 1896–1907.

OPPOSITE: *Paphiopedilum bellatulum* (as *Cypripedium bellatulum*) from Jean Jules Linden, Lucien Linden and Emile Rodigas, *Lindenia, Iconographie des orchidées*, 1885–1906.

Between 1996 and 2015, 1.1 billion live plants were traded, while 31 million kilograms of cut stems entered the market. Trade in wild plants is tightly regulated by CITES (the Convention on International Trade in Endangered Species) and the majority of orchids bought and sold in the world have been propagated responsibly. A market this valuable, where some prize rarity and novelty above all else, always risks falling foul of darker forces – and illicit trade in orchids is pushing some species to the brink of extinction.

The genus *Paphiopedilum*, sometimes known as slipper orchids, have a distribution across India, South-East Asia, China, New Guinea and the Solomon Islands. Flowers are exquisitely complex, with the lower labellum section of the flower resembling an elegant slipper. The purpose of this highly evolved structure is elegant too, facilitating the temporary capture of pollinating insects, holding them until pollen transfer is complete. The beauty this evolutionary advantage has brought *Paphiopedilum* now threatens the very existence of several species due to unscrupulous wild collection.

Paphiopedilum bellatulum requires a precise habitat, favouring the cracks and crevices of forest limestone outcrops – provided they're filled with suitable volumes of organic matter and receive constant local moisture. Distributed sporadically across south-west China, Myanmar and north-west Thailand, this species is under significant threat and rated "Endangered" on IUCN's Red List, with a little over 1,000 recorded plants left in the wild. Known as "the egg-in-the-nest" orchid, its curious flower shape and mottled petals, borne from a low rosette of spotted leaves, makes it highly distinctive and it's frequently plundered from the wild by illegal collectors. It's a frustratingly short-term piece of economics. While wild extraction is the quickest path to the (black) market, this approach has missed the opportunity to responsibly explore the genetic richness of the wild population to create new cultivars.

P. bellatulum's narrow ecological niche makes it inherently vulnerable to a changing climate or shifts in land management around it. At least one pressure has to be removed for this species to have chance of survival.

Anyone buying an orchid has a responsibility to ask where it has come from. Well-known online platforms list illegally collected orchids, in direct contravention of CITES, and buying a plant through this channel will imperil the survival of the species in the wild. Seeking clear and traceable evidence that an orchid has been propagated legitimately is one small but effective step to conserving this remarkable family of plants.

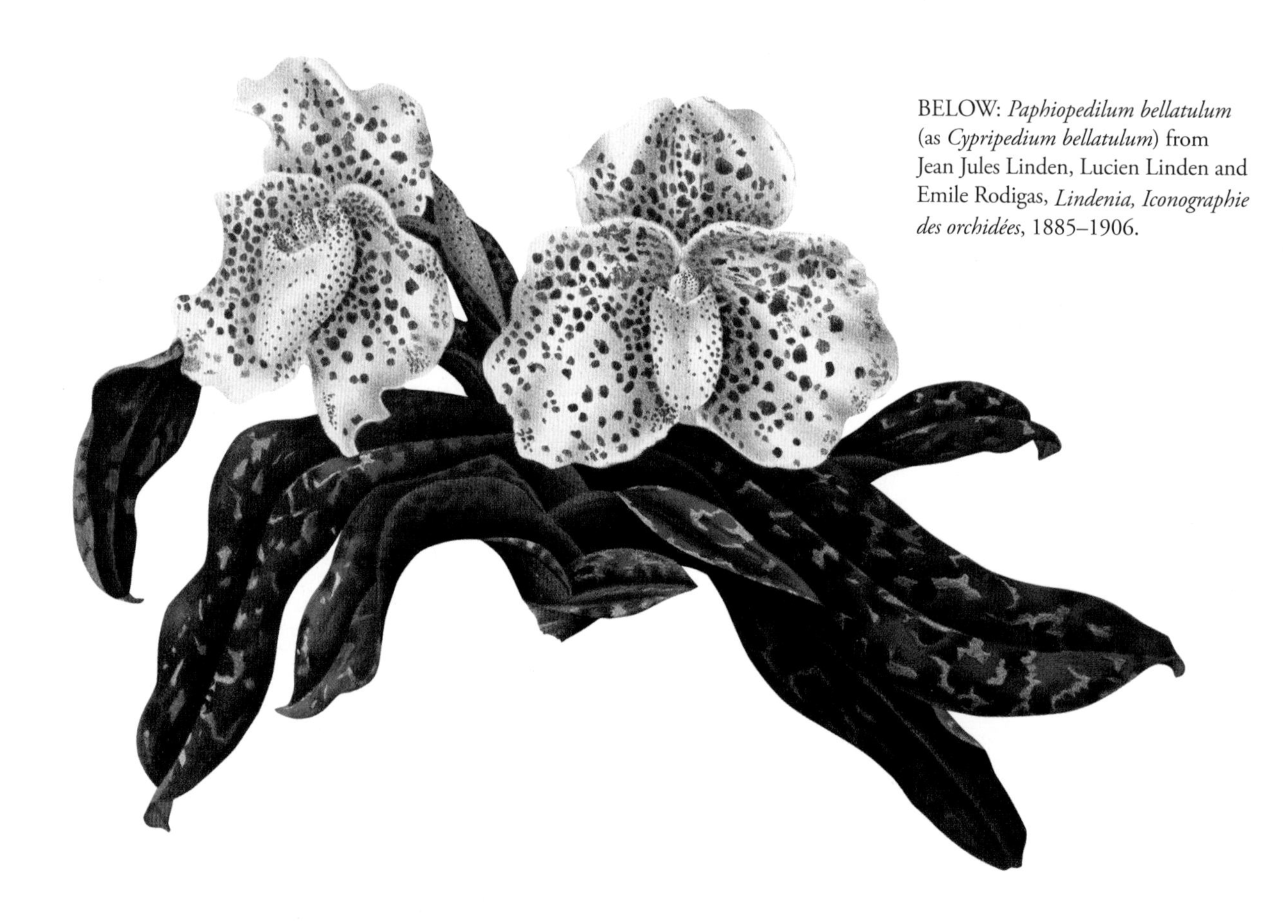

BELOW: *Paphiopedilum bellatulum* (as *Cypripedium bellatulum*) from Jean Jules Linden, Lucien Linden and Emile Rodigas, *Lindenia, Iconographie des orchidées*, 1885–1906.

Letter from Henry Alabaster, Royal Siamese Museum, Bangkok, Siam (Krung Thep, Thailand) to Sir Joseph Dalton Hooker, Director of Kew, 6 August 1884.

Alabaster was a British diplomat and royal advisor to the Thai King, as well as Director of the Royal Siamese Museum and Garden. He writes to inform Hooker that he has sent a Wardian case with several different plants including various orchids he found growing locally. He died on 8 August 1884, three days after writing this letter.

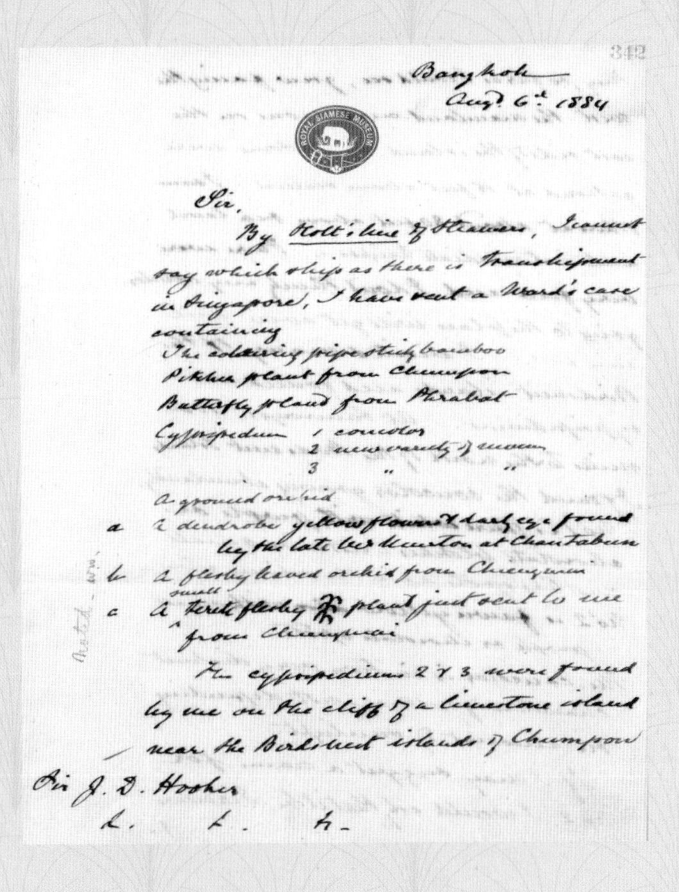

342

Bangkok
Aug. 6th 1884

ROYAL SIAMESE MUSEUM

Sir,

By Holt's line of Steamers, I cannot say which ships as there is transhipment in Singapore, I have sent a Ward's case containing
The soldiering pipe stick bamboo
Pitcher plant from Chumpon
Butterfly plant from Phrabat
Cypripedium 1 concolor
2 new variety of niveum
3 " "

A ground orchid
a A dendrobe yellow flowered & dark eye found by the late Mr Hunton at Chantabun
b A fleshy leaved orchid from Chiengmai
c A small terete fleshy plant just sent to me from Chiengmai

The cypripedium 2 & 3 were found by me on the cliffs of a limestone island near the Birdsnest islands of Chumpon

Sir J. D. Hooker
&c. &c. &c.

They, so far as I could see, grew facing the west - the mainland - and none on the east side of the island. Some were gathered at 15 feet above sea level - I saw others at about 80 feet above sea level and intermediate heights -. There were very few and I don't think any collector going to the place will get even one case full. I visited many of the Birdsnest islands and found no cypripediums. At Samroiyot 100 miles to the north of the Birdsnest Islands I found the concolor growing abundantly.
No 1 is pure white with purple or chocolate blotches - not mere small seed like spots as in the ordinary niveum
No 2 is fawn yellow ~~also~~ with small purple or chocolate spots.
The concolor is in two very distinct characters of yellow - not depending on amount of sunlight.

If I may suggest a name for No 2 I would say that it be Siamense &

343

I should have liked them to be named after the King and Queen, Chulalonkorn and Sawang, but I fear you may think those names ill sounding to European ears - All this is for you to decide.

I am sorry again to have failed in obtaining Gum Benjamin plants. Some were sent for me but never reached Bangkok, and now the survey party has been attacked by fever, several including one European have died, and all are in retreat. I shall have the honour of writing again about this when Mr McCarthy the head surveyor reaches Bangkok.

I have the honour to be
With very high respect
Your very faithful sert
H. Alabaster
Director Rl Garden

32

LONDON PLANE

Platanus × hispanica

Tall, slender, tough and tolerant, the London plane was made for the punishing urban environment. "Made", in this instance, is an appropriate term, for the London plane has never existed in the wild. The offspring of a transatlantic match between the Oriental plane (*Platanus orientalis*) and the American sycamore (*Platanus occidentalis*), the London plane came to be when these two species became acquainted in the garden of renowned seventeenth-century plant collector John Tradescant the Younger.

The London plane is known by several scientific names (a botanical challenge called synonymy) and you may see it referred to as *Platanus × acerifolia*, *Platanus × hispanica* or *Platanus hispida*. It's a handsome tree: finely proportioned and never thickset, with a silver and grey patchwork bark – a tessellation of geometric forms. Its large glossy maple-like leaves bear fluffy seed heads in late spring. Their tolerance of hard pruning has spawned an arboricultural art form, with skilfully crown-reduced trees looming over affluent city districts like woody giraffes.

John Tradescant the Younger could never have predicted quite what an urban superhero the London plane would become. City trees have been described as suffering simultaneous choking and dehydration; such is the brutality of the urban landscape. These trees are often growing in poor soil, surrounded by hard surfaces and radiated heat, and their restricted roots suffer regular disturbance. It's extraordinary any urban tree survives its first year, let alone becomes established and functional. Enlightened cities have developed more sympathetic environments, using permeable paving, "engineered" soil and more generous zones for roots, but heat, pollution and disturbance remain threats.

The London plane doesn't just survive in this challenging environment, it actually provides benefits for its (human and animal) fellow city dwellers. This hybrid tree, a botanical curiousity, is a remarkably effective pollution filter, contributing to the twenty-first century's most prized public good: clean urban air. The London plane absorbs atmospheric pollution, especially particulate matter, through its leaves and deposits it through the shedding of its plated bark. Walking through a park planted with London

OPPOSITE: *Platanus orientalis* from John Sibthorp, *Flora Graeca*, 1840.

ABOVE: *Platanus occidentalis* from Mark Catesby, *The natural history of Carolina, Florida, and the Bahama Islands*, 1754.

LEFT: *Platanus orientalis* by Auguste Faguet from Louis Figuier, *The Vegetable World*, 1867.

ABOVE: Herbarium specimen of *Platanus orientalis* collected by H. Lindberg in Cyprus in 1939, held at the Royal Botanic Gardens, Kew.

PLANTARVM

SEV

STIRPIVM

HISTORIA,

MATTHIAE DE LOBEL

INSVLANI.

Cui annexum est ADVERSARIORVM

VOLVMEN.

Reliqua sequens pagina indicabit.

ANTVERPIÆ,
Ex officina Christophori
Plantini Architypographi
Regij. M.D.LXXVI.

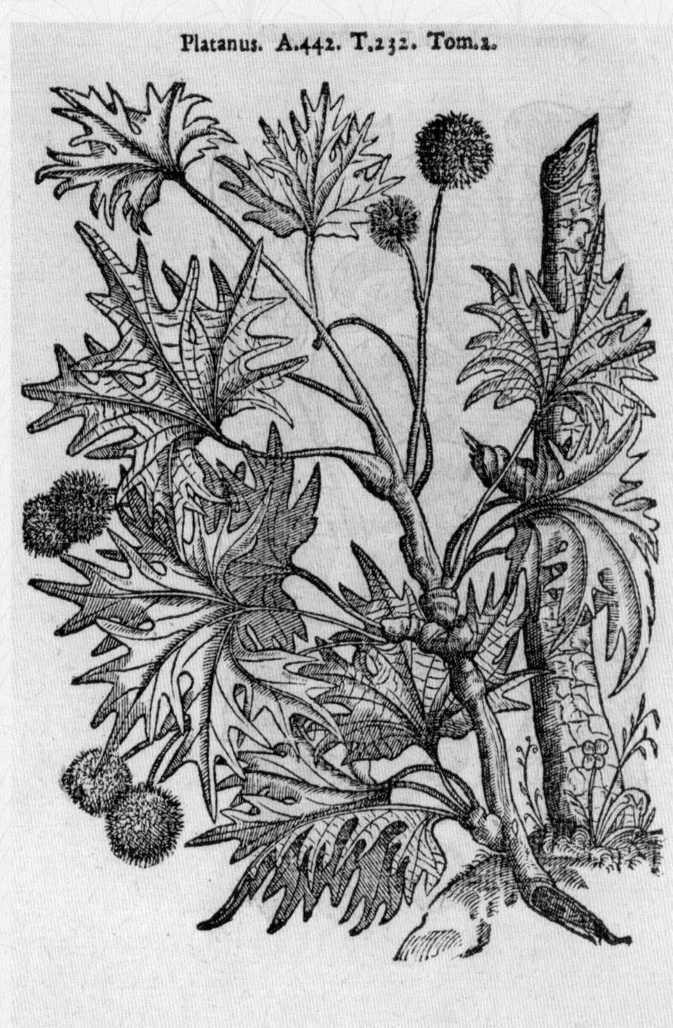

Title page and page illustrating *Platanus orientalis* from Matthias de L'Obel, *Plantarum seu Stirpium icones*, 1576.

This book is a combination of the illustrations published in the works of Dodoens, Charles de L' Ecluse and de L'Obel, arranged by L'Obel. The index is in Latin, Dutch, German, French, Italian, Spanish, Portuguese and English.

LEFT: *Platanus orientalis* by Mary Anne Stebbing, 1897. Kew Collection.

planes may be the healthiest choice a city pedestrian can consciously make.

The global population of London planes are genetically identical, cloned using vegetative (cuttings or grafts) methods to make new specimens. This monotony, combined with the challenge of thriving in a stressful environment, makes the London plane susceptible to new threats, especially novel pests and diseases. Our globally connected economy is moving these threats at an ever-accelerating rate. Beetles, moths, fungi and bacteria travel across continents on imported building materials or plants and become most unwelcome guests, finding new host species completely unadapted to the fresh danger they pose.

No species exists in isolation, and our native trees coexist with and survive a range of native menaces as well as foreign newcomers. Co-evolution with these known threats leads to a range of defences: producing antagonistic chemicals, such as volatile organic compounds to deter attack, and compartmentalizing fungal incursions to stop them spreading. Novel pests and diseases don't carry the same signatures to prompt defensive responses, or they attack London planes earlier in the year before their leaves have become unpalatably tough. The lack of predators to regulate the population of new threats allows proliferation in overwhelming numbers.

The London plane is not immune from this threat, and specimens around the world are under attack from new diseases. Massaria disease causes branches to drop; *Ceratocystis* fungus stains vital water-moving xylem tissue, causing severe wilting and death; *Phellinus* fungus causes white rot, digesting the structural lignin of the tree; and *Inonotus* fungus causes cankering, which leads to branch "failure" that splits the tree apart. The universal consistency of London plane trees means there's no resilience through variation, so one disease can cause equal havoc across a genetically identical population. The iconic London planes of France's Canal du Midi are succumbing to *Ceratocystis* at an alarming rate: over 15,000 trees have been removed since the disease took hold in the early 2000s.

Placing a proper value on the services street trees provide should justify the right investment to support this vital element of the urban landscape. A treescape diverse in species and genetics and grown in high-quality conditions – with aftercare that allows specimens not just to survive, but also to thrive – is an essential contribution to the health of a city.

33

PASQUE FLOWER
Pulsatilla vulgaris

For a few fleeting days every year, the pasque flower commands attention with its fragile beauty. In the rapidly lengthening spring evenings, its translucent petals glow atop a feathery mass of glaucous foliage. This compelling species is a subtle superstar plant.

The pasque flower is a member of the buttercup family (Ranunculaceae), reflected in its simple open structure and central mass of golden stamens. Its petals come in alluring shades of rich pink or purple, with a velvety outer surface. The seed head is as ornamental as the flower, a gossamer sea anemone that's an alluring conductor of light. Tradition associates its flowering with Good Friday (the word "pasque" is a derivation of "paschal", meaning "of Easter") and it reliably appears in mid-April, regardless of Julian calendar variations.

Pulsatilla vulgaris has a curious, scattered distribution across Europe, from the northerly reaches of Scandinavia through patches of the UK and highly selective spots in France, Switzerland, Romania, Poland and Slovakia. Its habitat requirements are precise. In the wild, it favours dry grassland with underlying chalk geology and it has a curious association with ancient history in the UK, with notable populations associated with Roman and Viking archaeological sites. Predictably, this association gave birth to the legend that pasque flowers sprang from the spilt blood of Viking warriors, but the reality is more prosaic. The undisturbed nature of these ancient sites ensures the stable conditions that a slow-to-establish plant like the pasque flower needs. Those elegant seed heads bear few viable seeds, so large wild populations of pasque flower are testament to a stable long-term habitat.

In cultivation, *P. vulgaris* has traditionally been treated with the precision its delicate nature suggests. The fine art of alpine gardening, with precise manipulation of growing media to ensure exacting pH levels, perfect drainage and invigorating ventilation, has replicated wild conditions and presented thriving pasque flowers to rock garden visitors. However, recent trends in horticulture have created a new role for this flower. Grown as a component in complex designed plant communities ("ecological horticulture"), *P. vulgaris* shrugs off its fragile reputation and becomes a

ABOVE: *Pulsatilla vulgaris* (as *Anemone pulsatilla*) from Georg Christian Oeder *et al., Flora Danica*, 1761–1883.

OPPOSITE: *Pulsatilla vulgaris* (as *Anemone pulsatilla*) from Edward Hamilton, *Flora Homoeopathica*, 1852.

a
b
c
d
e

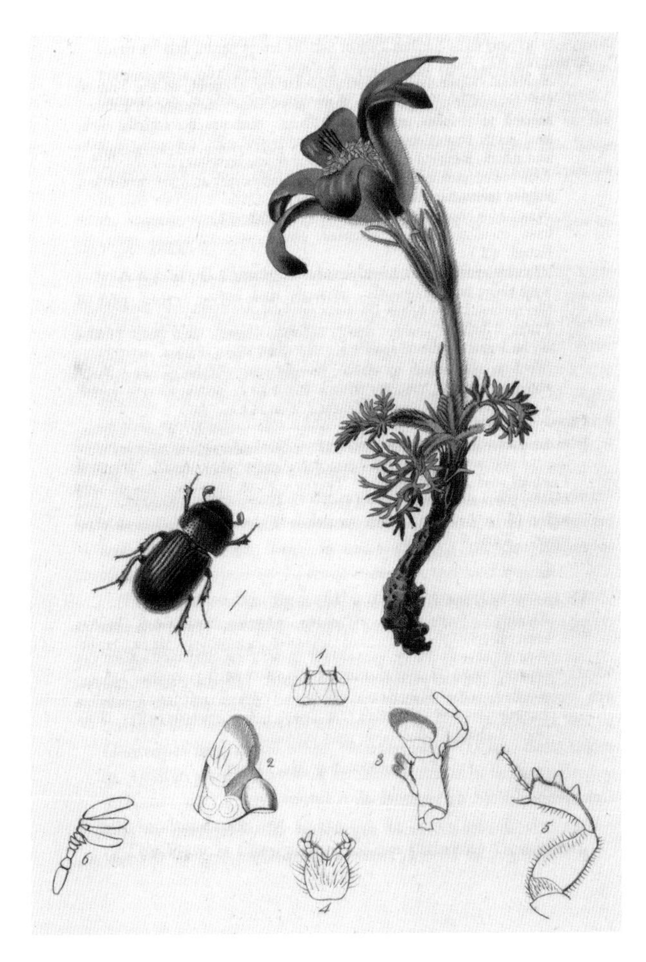

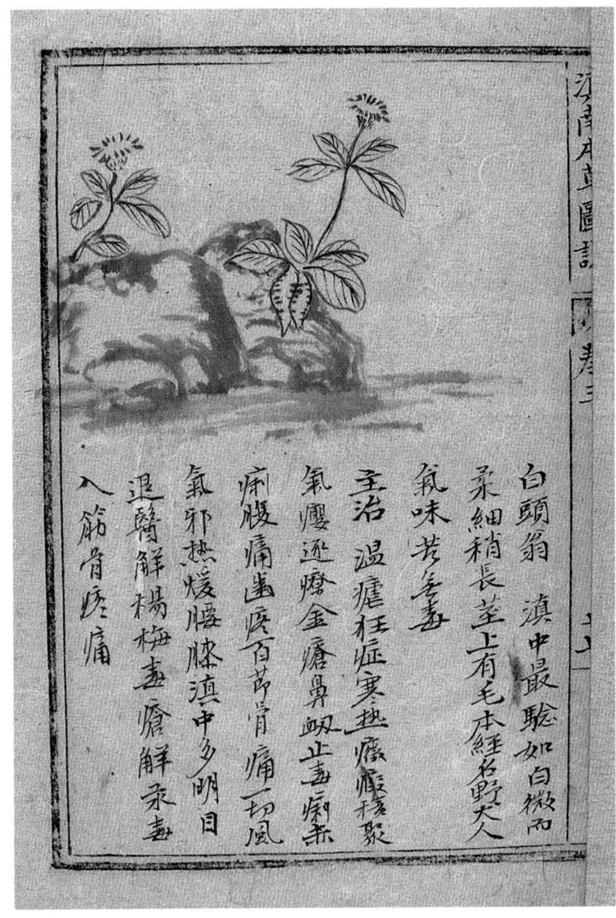

robust player, forming strong rosettes among grasses and later, flowering forbs. In this new context, it's clearly capable of finding a complementary niche above and below ground, growing in harmony with new neighbours.

The beauty of the pasque flower has not helped its fortunes and its precise ecological requirements don't make it a resilient plant in a rapidly changing world. Populations across Europe have declined significantly in the last century and IUCN now classify it as "Near Threatened". Low-intensity sheep grazing shaped the ideal niche for the pasque flower, providing closely cropped, nutrient-poor grassland where a consistently stressed environment stopped competitive plants from subsuming others. Chalk grassland is a plant community in a state of arrested succession. Without the stress and disturbance of grazing, it will revert first to wiry scrub and then woodland.

There's a curious irony that grazing – a tool our species has harnessed for thousands of years – drives biodiversity in some situations and becomes a destructive force in others. Goats introduced to an island rife with endemic plants can force species into extinction, while carefully managed herds and flocks on species-rich grassland create equalized environments for many species to thrive. Species recovery plans for the pasque flower recognize the complex requirements of chalk grassland plants and the "just right" levels of grazing needed to maintain a favourable habitat.

Should we intervene at all? The concept of re-wilding is sometimes misread as simply walking away from our landscapes and allowing nature to do the rest. Many of Britain's most biodiverse landscapes, from heathland to chalk grassland, have developed through human stewardship. These landscapes support a rich array of fauna, and the fate of these diverse, complex habitats is in our hands.

OPPOSITE: *Pulsatilla vulgaris* (as *Anemone pulsatilla*) from Dieterich Leonhard Oskamp, *Afbeeldingen der Artseny-Gewassen met Derzelver Nederduitsche en Latynsche Beschryvingen*, 1800.

ABOVE LEFT: *Pulsatilla vulgaris* (as *Anemone pulsatilla*) from John Curtis, *British Entomology*, 1823-40.

ABOVE RIGHT: Ming herbal painting of Chinese pulsatilla root from Lan Mao *Diannan bencao tushuo* (*The Illustrated Yunnan Pharmacopoeia*), *c.* thirteenth/fourteenth century. Wellcome Collection.

XXI, 5.
34. Cupuliferae.
A
B
1
2
3a
3b
4
5
6
7
W.MÜLLER, GERA.
161. Quercus pedunculata Ehrhart. Stieleiche.

34

COMMON OAK

Quercus robur

Great Britain's historic naval power was founded on the strength and endurance of the oak, a tree planted in abundance during the Elizabethan era to ensure generations of future shipbuilding material. The timber, with its true grain and rich tannin content, had immense value beyond its maritime applications, and underpins centuries of construction.

The oak doesn't just make dependable homes for us: the oak is Britain's most hospitable tree, with its fissured bark and familiar lobed leaves hosting hundreds of species of invertebrates, birds, mammals, fungi and plants.

Mature oaks have one of the more recognizable forms in the British landscape, their short trunks and fan-like canopies becoming increasingly striking with age. In woodlands, they support a diverse understory of hazel, holly, anemones and ferns, in contrast to fellow climax tree the beech, which engineers an empty woodland floor through a competitive mechanism called allelopathy. In fields and parklands, the oak becomes a sentinel, forming iconic profiles that catch the eye from afar.

The oak is capable of reaching extraordinary old age, with examples such as Lincolnshire's Bowthorpe Oak estimated to be over 1,000 years old. An old country saying depicts oaks as spending "300 years living, 300 years resting and 300 years declining", and veteran trees take on an remarkable form: thickset, hollow-trunked and fissured, with modest sprigs of live growth garlanding the stubby extremities of the tree.

Great Britain has two species of wild oak: the pedunculate (*Quercus robur*), which grows on lowlands, primarily in southern England, and the sessile (*Quercus petraea*), which is found on higher ground, especially moorland, where it forms wizened, wind-flagged specimens.

The longevity and strength of the oak has inspired countless examples of myth, symbolism and allegory from Zeus to the National Trust. Veteran UK oak trees are associated with giants (Gog and Magog), the English Civil War (several oaks claim to have hidden Charles I) and Robin Hood. In Europe, the oak aligns to gods pagan (Thor in Scandinavia) and Christian.

OPPOSITE: *Quercus robur* (as *Quercus pedunculata*) from Otto Wilhelm Thomé, *Flora von Deutschland*, 1886–89.

ABOVE: *Quercus robur* from Elizabeth Blackwell, *Herbarium Blackwellianum*, 1765.

OPPOSITE: *Quercus robur* from Dieterich Leonhard Oskamp, *Afbeeldingen der Artseny-Gewassen met Derzelver Nederduitsche en Latynsche Beschryvingen*, 1800.

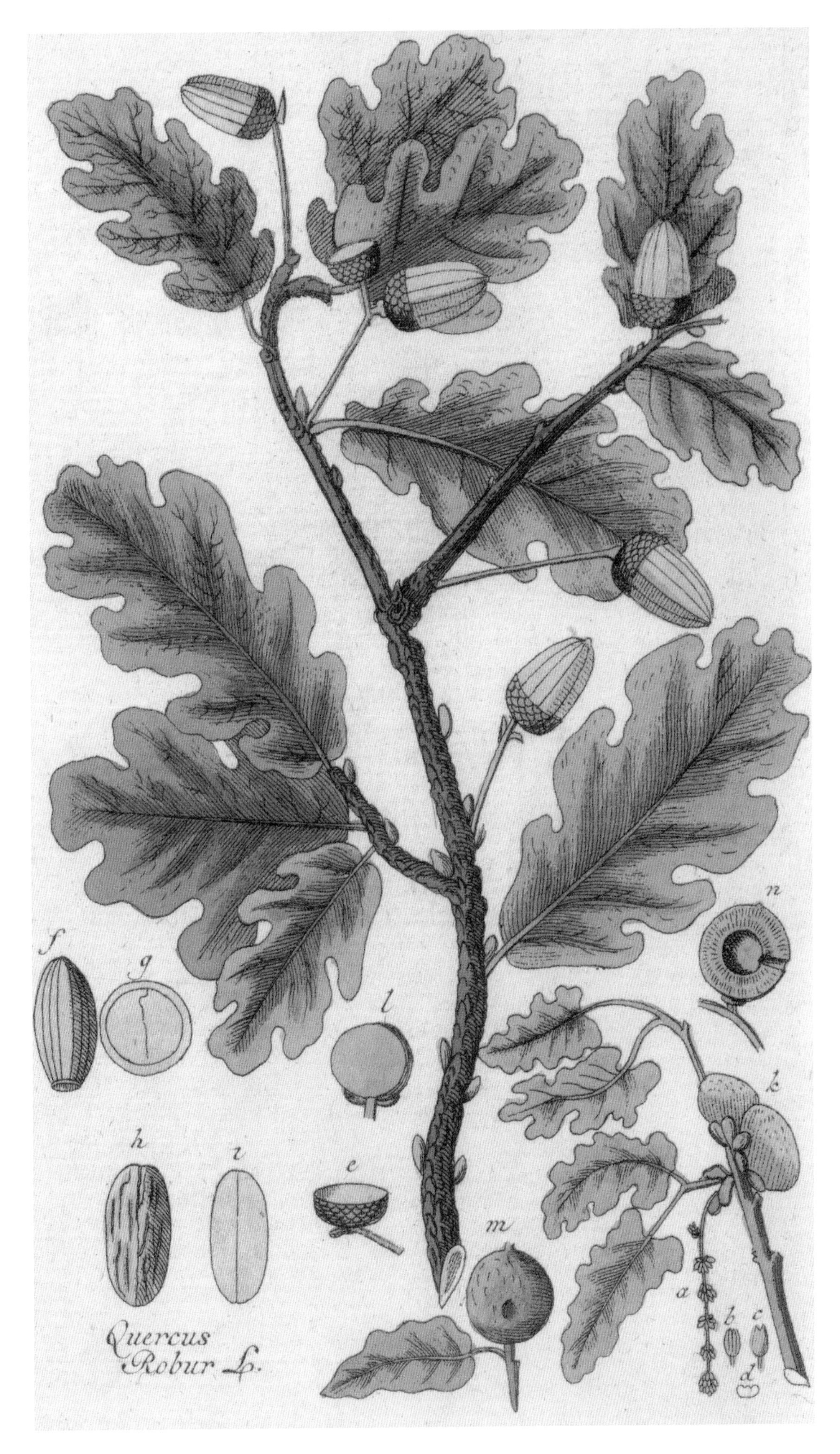
Quercus
Robur L.
f
g
l
n
k
h
i
e
m
a
b
c
d

Beyond the value of the exceptional timber, the tannin-rich bark is used to tan leather, while a diet rich in the tree's acorns fattens the famous Ibérico pigs of Spain before they become expensive ham. While screw caps and plastic stoppers are growing in popularity, some of the best wine is still corked with the spongey bark of the cork oak (*Quercus suber*).

The oak tree's long life endows it with a range of survival strategies. As trees become veterans, their canopy slowly diminishes, reducing their resource requirements. Decay can be compartmentalized to reduce its spread, and many fungal pathogens can co-exist within the tree for centuries. Despite these resilient strategies, even the oak cannot withstand the accelerating pressure of global plant threats, and it's deeply unsettling to see such a robust species declining rapidly in our cities, towns, countryside and woodlands.

A highly effective array of pests, diseases and environmental pressures are aligning against the oak. The causes of Acute Oak Decline are still being researched, but are believed to result from woodboring beetles transmitting a novel bacteria capable of overcoming the oak's natural resistance. The results are distressing: lesions, stem bleeding and a rapid decline in condition, with death occurring as quickly as five years after infection. Equally worrying is the complex condition Chronic Oak Dieback, believed to be caused by a cocktail of issues, including waterlogged soil, pathogens, climate pressures and pollution.

The caterpillars of the Oak Processionary Moth, an imported arrival from southern Europe, can rapidly denude canopies of mature trees, reducing vigour and undermining the tree's long-term health. In the UK, the Oak Processionary Moth is currently trapped within the M25 orbital motorway that surrounds London, but it's only a matter of time before it spreads further; can the mighty oaks of the British countryside withstand yet another threat? Let us hope that with our help the magnificent oak will provide myths and shelter for generations to come.

LEFT: *Quercus robur* subsp. *robur* (as *Quercus fastigiata*) by Pancrace Bessa from H. L. Duhamel du Monceau, *Traité des arbres et arbustes, Nouvelle édition*, 1819.

OPPOSITE: *Quercus robur* (as *Quercus pedunculata*) from *The Garden*, 1873. This article describes the chapel-oak of Allouville-Bellefosse in France, said to be 1,000 years old and containing two small chapels, one on top of the other.

Elder, planted freely. As cover in woods and plantations, where little else would live, keepers used, in winter, to dibble in cuttings of Elder in all bare, naked places, being well aware of its utility as a plant for "thickening up." Lastly, the Elder makes a good plant for filling up gaps in hedges, especially where they pass under trees, and for boundary fences, where nothing else will grow. It will preserve the continuity of a hedge right up to the trunks or stems of even Beech and Horse Chestnut.—THOS. WILLIAMS, *Bath Lodge, Ormskirk.*

THE CHAPEL-OAK OF ALLOUVILLE (PAYS DE CAUX).

THIS very remarkable tree is a specimen of the common Oak (Quercus pedunculata), which is believed to be about 1000 years old, and is the object of a considerable amount of veneration to the inhabitants of the district, from the circumstance that its hollow truuk has long been used as a chapel. Properly speaking, it contains two chapels, one above the other, access to the upper one being obtained by means of a spiral staircase on the outside, as shown in the illustration. Over the entrance into the lower chapel is the following inscription : " A Notre-Dame-de-la-Paix, érigée par M. l'abbé du Detroit, en 1696."

The Chapel-Oak of Allouville.

The tree is now about 50 feet high, its top having been broken by the wind, or cut off, at some remote period, and in its place a sort of bell-tower has been erected, the top of which is about on a level with the highest branches. The lower part of the trunk is more than 11 feet in diameter. The bark, which is of a corky nature and deeply fissured, is upwards of 4 inches in thickness. Although so very old and hollow, the tree still exhibits a vigorous growth, its huge branches, which extend over an area of 2478 square yards, being annually covered with an abundant foliage, and usually bearing a large quantity of acorns. The extraordinary age of this venerable tree renders it impossible to obtain any particulars of its early history. Local tradition, however states that the district was formerly covered by a natural Oak forest, of which the Chapel-Oak of Allouville is now the sole survivor.

THE INDOOR GARDEN.

CHINESE PRIMROSES.

THE Chinese Primroses range in colour from the deepest purple to the purest white. They are improving so fast that it is impossible to specify size with exactness. The largest sort would perhaps cover a crown-piece ; they are fringed, serrated, marbled, and ringed, in the most various and beautiful manner; stem after stem, heavily crowned with masses of flowers, rises boldly and to a goodly stature out of their hearts—crowns of glory supported by the beautifully-formed fern-shaped or other leaves. There are double varieties of purest white to purple, with many intermediate shades, forms, and sizes. These are more lasting and useful, though not more beautiful, than the single varieties ; but they are invaluable for cutting for vase or bouquet work, which the single ones are not, the flower soon separating from its green calyx when cut. The double will stand a week, a fortnight, three weeks, or more in water, fresh and sound as at first ; and for bouquets, the Double White rivals in usefulness the Camellia, Stephanotis, or Gardenia. Scarcely any plants can be easier grown than these Primroses. Properly treated, their natural season of flowering may be said to be from November to April. The sun, so essential to most flowers, may be said to be unfavourable to these ; as he gains strength they, unless shaded, lose beauty and freshness. But in winter and early spring they glow with a beauty, shine with a brilliancy and purity of colouring, almost unequalled. All the single varieties are best treated as annuals. After flowering, throw the plants away. Seed saving needs special skill in selection and management, and the amateur or lady gardener, unless an enthusiast, had better not attempt it. Any respectable nurseryman or florist will supply good seeds. Sow in light soil—peat or loam with a fourth part of sand—in February or March ; cover the seed with an eighth of an inch of soil, and keep it rather dry until it begins to grow. As soon as the plants can be handled, prick them out, about six round

35

AFRICAN VIOLET

Streptocarpus ionanthus

The African violet starts many lifelong relationships between people and plants. These days, it is available in an enticing range of colours in garden centres, superstores and market stalls across the world, its willingness to flower copiously from a compact rosette of purple-veined leaves making it a highly desirable houseplant. Despite this ease of availability, it's not the easiest plant to grow at home, needing regular watering, some humidity and plenty of indirect sunlight. Despite the heavily domesticated appearance of the African violets, these precise environmental requirements are a direct consequence of the plant's wild origins in the sacred forests of coastal Kenya and Tanzania. It thrives in primary (undisturbed) forests, growing in a precise niche: limestone rocky outcrops with diffuse light through the tree canopy.

The plants were originally given the genus name *Saintpaulia*, a name referring to Baron Walter von Saint Paul-Illaire, a colonial governor of Tanganyika (now a region of Tanzania), who discovered them at the end of the nineteenth century. Following some recent taxonomic work, the genus *Saintpaulia* has now been absorbed into the genus name *Streptocarpus*. Several other closely related species of African violet are found in Tanzania and Kenya, including the elusive *Streptocarpus teitensis*, which is endemic to only one location among the Teita Hills in southern Kenya.

LEFT: *Streptocarpus ionanthus* (as *Saintpaulia ionantha*) from Eduard Regel, *Gartenflora*, 1893.

OPPOSITE: *Streptocarpus ionanthus* (as *Saintpaulia ionantha*) by Matilda Smith from *Curtis's Botanical Magazine*, 1895.

OPPOSITE: *Streptocarpus ionanthus* (as *Saintpaulia ionantha*) from *Revue horticole*, 1902.

RIGHT: Herbarium specimen of *Saintpaulia ionantha* collected in 1894, held at the Royal Botanic Gardens, Kew.

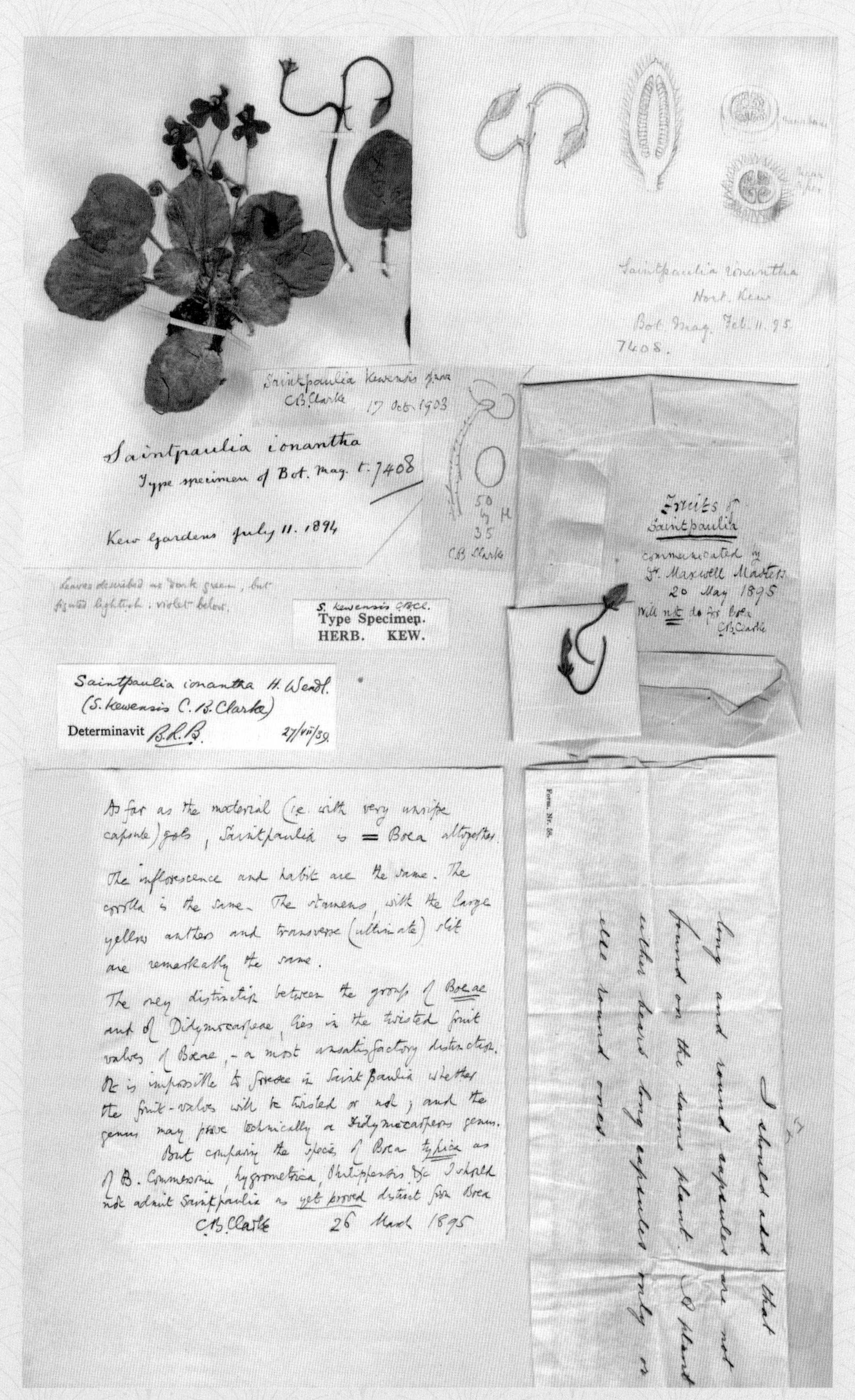

Taxonomy, or the science of classification, has been transformed by the arrival of DNA sequencing. Until the 1980s, plants were identified by their morphology or physical appearance, with hand lenses and microscopes employed to discern minute differences between species. Then a revolution occurred. By analyzing and comparing DNA sequences between species, a dazzling new perspective on the tree of life was possible. The very fabric of plant classification was disrupted with orders, families, genera and species split, joined and clumped into new relationships.

Saintpaulia is an interesting example of the fascinating changes molecular sequencing has generated. Now correctly referred to as *Streptocarpus ionanthus*, our African violets are highly adaptable to their environment, showing a range of physical forms in a trait known as "plasticity". DNA sequencing has now revealed eight subspecies of a plant we originally thought was one variable species. The new classification was not a moment of triumph, for with it came the realization that the plant was growing in fragmented, greatly diminished populations and it became classified as "Near Threatened" on IUCN's Red List.

There's a striking disconnect between the multimillion dollar turnover of the African violet houseplant industry and the economic forces driving deforestation and the subsequent decline of its wild ancestor. The genetic resources that create this most desirable of living home accessories were taken from their native land without payment or royalty agreement, so no financial benefits are reaped in Tanzania or Kenya from this bounteous crop.

The Convention on Biological Diversity spawned the Access and Benefit Sharing Agreement (sometimes called the Nagoya Protocol) to ensure the countries with rich genetic resources share the full benefit of them becoming monetized. This framework drives Kew's approach to conservation overseas, with scientists such as Africa Coordinator, Tim Pearce, defining new projects around the stated benefit of the partner country. It shifts the focus of a botanic garden from acquisition to provision, creating seed bank resources in-country and supporting regional partners in the classification and conservation of biodiversity.

ABOVE: *Streptocarpus* x *kewensis* from *L' Illustration horticole*, 1854.

OPPOSITE: *South African Flowers, and Snake-headed Caterpillars*, by Marianne North, 1882.

Lucy T Smith 200

SUICIDE PALM

Tahina spectabilis

Madagascar is a biodiversity hotspot, one of eight global centres of exceptional endemism. The island, the world's fourth largest (over 587,000 square kilometers), has over 12,000 recorded plant species, a staggering 89 per cent of them endemic. The diversity of the island's fauna is even more remarkable, with 98 per cent of Madagascar's amphibians, reptiles and mammals unique to its shores, including the iconic lemur.

A wealth of factors shaped the abundance of flora and fauna on this remarkable island. Long-term isolation from neighbouring continents, estimated at 88 million years, allows substantial time for speciation. The diversity of ancestors that first populated the new island is significant too. The formation of Madagascar started in the centre of ancient supercontinent Gondwana. As Gondwana broke apart, Madagascar left the land mass that became Africa, becoming attached to the proto-Indian subcontinent, breaking away before it started its collision course with the Himalayas. This melting pot of flora and fauna, left alone in the midst of the Indian Ocean, was a potent starting point for biodiversity.

Partly due to its varied geological origins, Madagascar is an island of immense physical diversity as well. Arid, tropical and temperate, with mountains, spiny forests and rainforests, Madagascar has classified seven distinct ecoregions, defined by sharp contrasts in climate, altitude, soil and vegetation type. Given Madagascar's size, diversity and complexity, it's not surprising to learn new species are being discovered with regularity. Between 1999 and 2010, 615 discoveries were made, including a remarkable new plant that was seemingly hiding in plain sight.

During a walk on Ampasindava Peninsula in 2006, local cashew plantation owner Xavier Metz noticed an unusual plant. Although not a professional botanist, his affinity with the local flora caused this distinctive specimen to stand out. Not only was it novel, it was huge, a palm tree of imposing dimensions. Photographs sent to Kew, a global centre for palm expertise, quickly stirred excitement: it was either a mysterious outlier of the Asian genus *Corypha* or something new to science. A field trip was rapidly assembled, DNA samples and herbarium specimens gathered,

OPPOSITE: *Tahina spectabilis* by Lucy Smith, 2009. © Edward Twiddy.

ABOVE: Herbarium specimen of *Tahina spectabilis* collected by Mijoro Rakotoarinivo in Madagascar in 2007, held at the Royal Botanic Gardens, Kew.

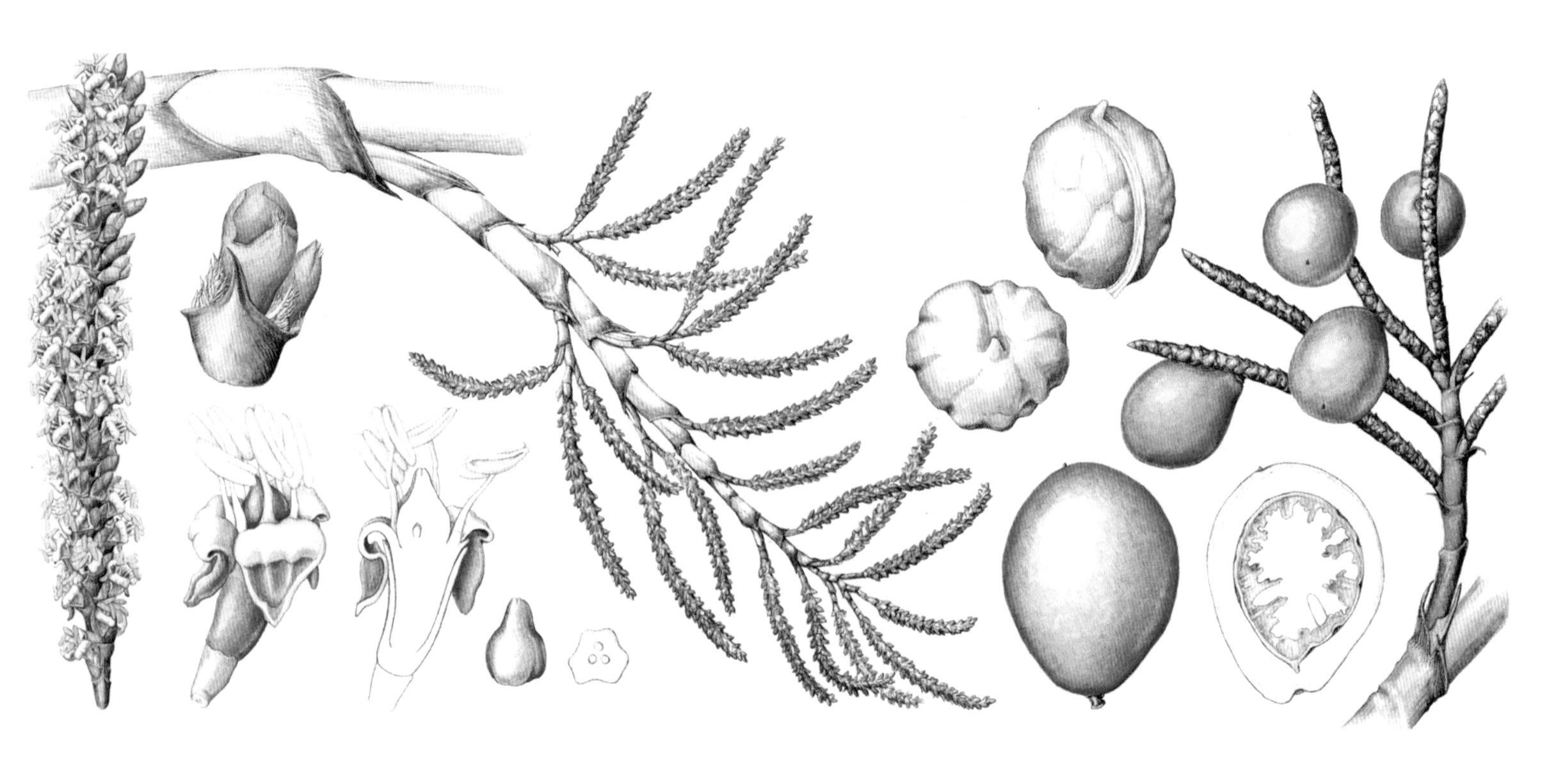

and comparisons against "type" specimens made, which confirmed the exciting hypothesis: a palm new to science.

Tahina spectabilis is aptly named. Standing over 18 metres tall at maturity, with leaves 5 metres wide, its discovery instantly made it Madagascar's largest palm. Size is not its most remarkable feature, however. Every 30 to 50 years, *T. spectabilis* forms an extraordinary pyramid-shaped terminal inflorescence, produces copious volumes of seed and dies, its hereditary commitments complete. This phenomenon, known as hapaxanthy, may be the reason why this palm had not previously been identified, its most distinctive feature potentially not revealed for a generation.

Hapaxanthy is not a trait exclusive to *T. spectabilis*. It's seen in other palm species and certain bamboos. Bamboos flowering and seeding can cause ecological shockwaves. Referred to as "mautam" (bamboo death), the abundance of food caused by massed flowerings in northern India induces a population explosion of the black rat. The subsequent plague hoovers up the bamboo seed and turns its attentions to the human population's food stores, threatening famine.

Despite hapaxanthy not being exclusive to *Tahina*, this exciting discovery was quickly and melodramatically dubbed "the suicide palm" by a fevered press. The novel find caught the mainstream imagination and was placed in the Top Ten species discoveries for 2008 (the year it was formally verified) by the International Institute for Species Exploration (IISE).

With the confirmation of a species new to science came the parallel discovery that there were hardly any *T. spectabilis* palms left. Initially, only 29 trunked adult specimens were counted, rendering it "Critically Endangered" on IUCN's Red List. Subsequent expeditions to Madagascar have found additional populations, although the palm remains perilously close to the brink, with just 50 adult specimens currently recorded.

Carefully harvested seed supplies a regulated propagation programme, bringing *T. spectabilis* into cultivation in botanic gardens around the world and forming the potential for reintroduction (albeit from a narrow gene pool). Under the auspices of an access- and benefit-sharing agreement commercial revenue from this new horticultural trade is returning to where it belongs: the communities where *T. spectabilis* grows.

ABOVE: *Tahina spectabilis* by Lucy Smith, 2009. © Edward Twiddy.

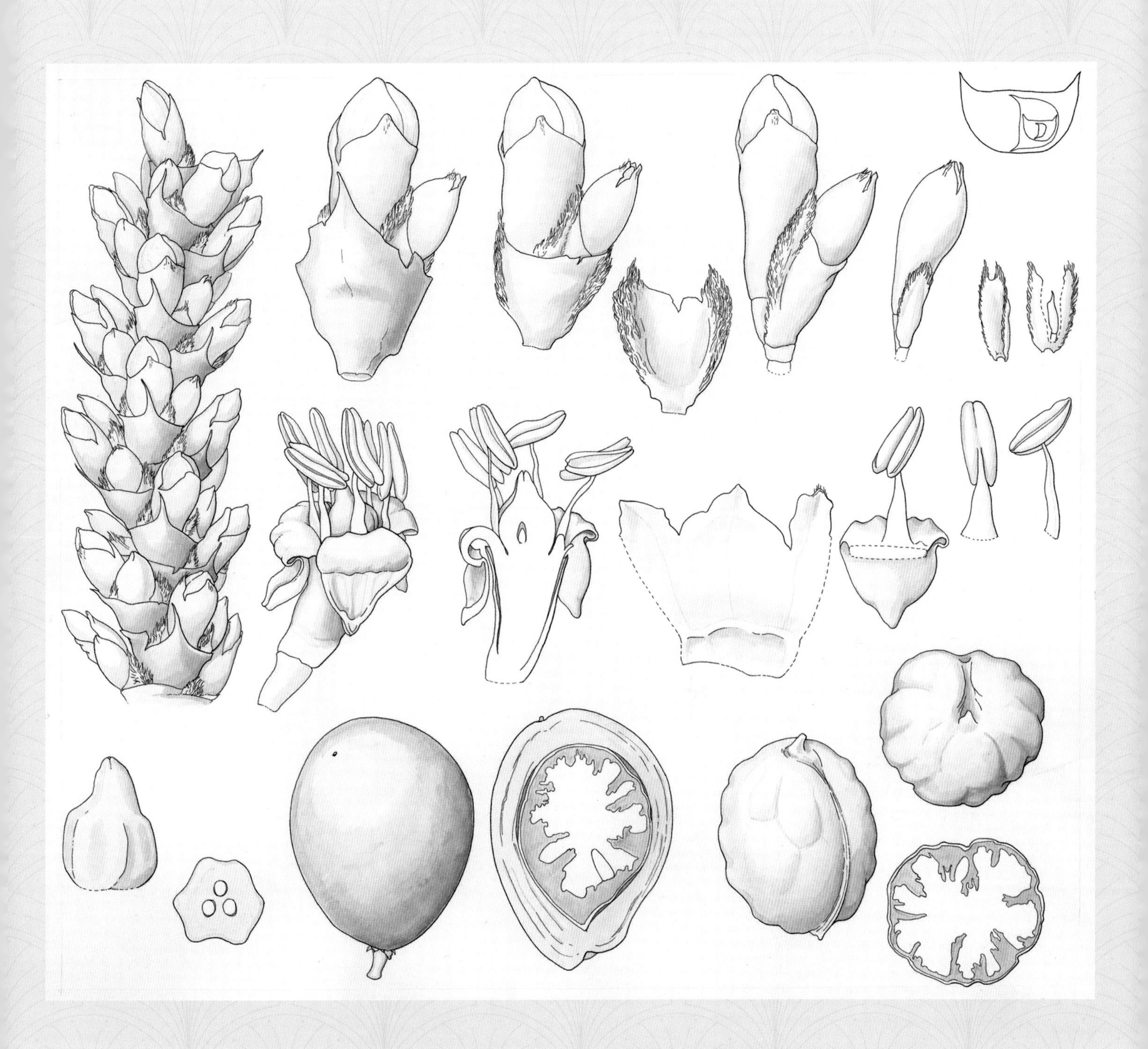

ABOVE: Flower and fruit material of *Tahina spectabilis* by Lucy Smith from John Dransfield *et al.*, *Genera Palmarum*, 2008.

37

CHILEAN BLUE CROCUS
Tecophilaea cyanocrocus

There a few plants with a more vivid colour than the Chilean blue crocus (*Tecophilaea cyanocrocus*). An electric blue, seemingly possessed of its own luminescence, this beautiful cormous plant has been a prized acquisition for the alpine gardener for over a century. The electrifying beauty was thought to be extinct in the wild, another lamentable tale of loss until, miraculously, it was rediscovered in a rare role reversal: a species found, not lost.

A diminutive 10 centimetres tall, the petite dimensions of the *T. cyanocrocus* lend themselves well to that most exacting of horticultural pursuits, the rock garden. In this precise environment, small but compelling specimens are revered for their fragile beauty, moved from the background supply to display glasshouse while flowering and returned before dowdiness sets in. The Gentian-like blue of the flower is the dominant feature of *Tecophilaea* and comprises a third of the plant, not including a couple of narrow green leaves at the stem's base.

T. cyanocrocus is a corm-forming (a bulb-like underground structure) plant from the Chilean Andes. It can be found at altitudes between 2,000 to 3,000 metres, growing above the tree line in stony scree, its wondrous colour a ravishing enrichment of an otherwise harsh environment.

The skill of the horticulturist makes plants from the high Andes at home in a suburban glasshouse. A crucial skill in making alpine plants thrive in cultivation is knowing not what to add, but what to take away. Impoverished stony soil and a harsh mountain climate have shaped this plant's physiology, and damp roots or humid air will quickly hasten a decline. Rock gardens or alpine houses with bright, direct sunlight, brisk ventilation to replicate the Andean atmosphere and sharp drainage around the roots are essential elements for horticultural success.

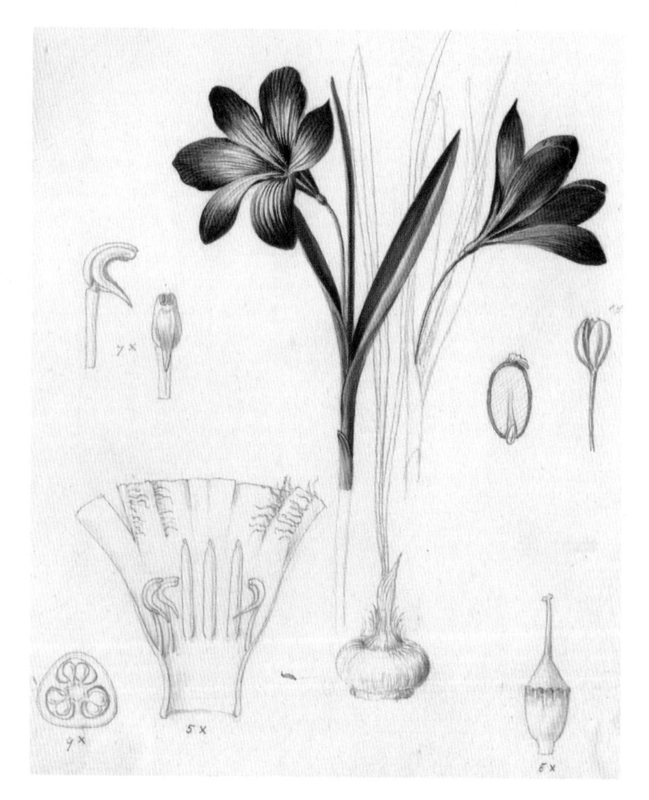

ABOVE: *Tecophilaea cyanocrocus* by Lillian Snelling from *Curtis's Botanical Magazine*, 1923.

OPPOSITE: *Tecophilaea cyanocrocus* from *Revue horticole*, 1900.

LEFT: *Tecophilaea cyanocrocus* from Eduard Regel, *Gartenflora*, 1872.

ABOVE: *Tecophilaea cyanocrocus* from Eduard Regel, *Gartenflora*, 1853.

Alpine gardening books written before 2001 are clear that *T. cyanocrocus* is extinct in the wild, a status first declared in the 1950s. It's an indication of less-enlightened times that one of the main drivers for extinction – exploitative over-harvesting of corms for the horticultural trade – doesn't always provoke contrition in these tomes. While many bulbs, corms and tubers sold are from sustainably cultivated sources, wild removal (usually on an industrial scale) is still an issue. Though it's often a vital source of income for those who are diminishing the wild population, the imperilled species suffers a heavy cost in exchange.

The assumption that *T. cyanocrocus* only existed in horticulture was shattered in the spring of 2001, when an extensive new population was discovered on privately owned land near Chile's capital, Santiago. While this was a welcome discovery, the shifting of *Tecophilaea*'s IUCN rating of "Extinct in the Wild" (and inclusion on the Red List) to an only marginally less severe "Critically Endangered" indicates many of this fragile species' threats are ever-present. One vital outcome from this discovery is a chance to expand the genetic diversity of the ex-situ conservation collections held in botanic gardens such as Kew.

Prior to the discovery of the new wild populations, the pre-2001 collections had been genetically sequenced to assess their potential for reintroduction to the wild. The results were disappointing, revealing virtually no diversity across the population held in cultivation. This would form a poor resource for any potential reintroduction, with little resilience to future threat. Without the individual resistance genetic variability brings – which can then be conferred onto future generations – new pathogens could uniformly infect and destroy a population.

Could other extinct species be found anew in the wild? Ecologists now have sophisticated remote surveying methods, using drones or satellites to assess large swathes of land in remarkable detail, increasing the chances of discovering isolated or relict populations. Private land ownership, inaccessible environments, political instability and simply the vast scale of some of our most biodiverse habitats mean that the chance of surprise discoveries remains ever-present.

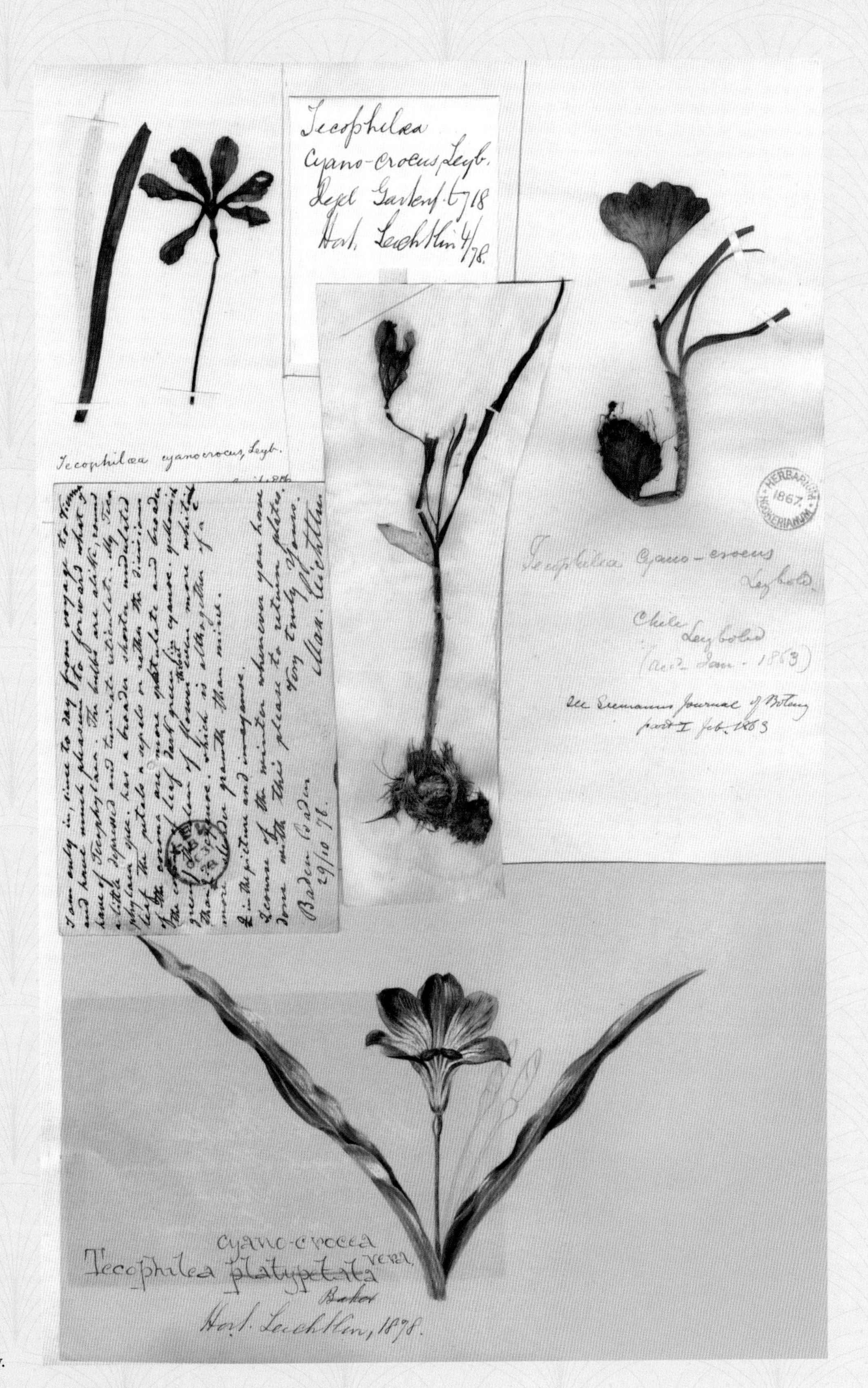

Herbarium specimen of *Tecophilaea cyanocrocus* collected by Leybold in 1863, held at the Royal Botanic Gardens, Kew.

ST HELENA EBONY

Trochetiopsis ebenus

The island of St Helena is a notably remote location comprising just 122 square kilometres of land in the southern Atlantic. Prior to the opening of its infamously windy airport in 2016, the British Overseas Territory could only be reached via a 10-day voyage from Cape Town, South Africa.

When the British sought to exile Napoleon and his imperial ambitions as far away as possible, St Helena was the ideal location – the ship transporting him was at sea for 12 weeks – but the chief beneficiary of this sort of geographic isolation is biodiversity. A wide range of influences drives speciation (the creation of distinct species). Climate, soil, topography, land use and ecological interactions can help determine the divergence of genetically and morphologically distinct species away from common ancestors.

St Helena is a volcanic island, topographically and environmentally diverse and nurtured by a benevolent climate. Its flora and fauna have evolved in total isolation for millions of years, leading to a significant amount of endemism: plants and animals that occur nowhere else in the world. Over 500 unique species have been recorded on the 47 square miles of St Helena, including 50 plants. But the island is no longer a pristine wilderness. The condition of this extraordinary biosphere reflects the truth of the pressures faced by biodiverse islands worldwide, no matter how remote: the influx of people and their imported, exotic species has a rapidly deleterious effect on biodiversity.

St Helena was settled in the 1500s. With people came their grazing animals, agricultural crops and ornamental plants – companions familiar to the settlers, but utterly alien to St Helena's ecological equilibrium. Plants that hadn't

OPPOSITE: *Trochetiopsis erythroxylon* (as *Pentapetes erythroxylon*) by Sydenham Teast Edwards from *Curtis's Botanical Magazine*, 1807.

BELOW: *Views of St Helena* by William Burchell from his *Saint Helena Journal*, 1806–10.

ABOVE: Herbarium specimen of *Trochetiopsis ebenus* collected by William Burchell in the early nineteenth century in St Helena, held at the Royal Botanic Gardens, Kew.

OPPOSITE: *Trochetiopsis melanoxylon* (as *Melhania melanoxylon*) by J. N. Fitch from John Charles Melliss *St. Helena: A Physical, Historical, and Topographical Description of the Island*, 1875.

evolved with generalist grazers (primarily goats) weren't unpalatable enough to deter munching. The pioneering seeding and colonizing methods of introduced plants were more efficient and less inhibited than resident flora. The newcomers quickly colonized space by out-performing endemic species.

The genus *Trochetiopsis* is unique to St Helena, and reviewing its conservation status is a dispiriting experience. Three species are known to science: *T. ebenus* (St Helena ebony) is rated "Critically Endangered" by IUCN, *Trochetiopsis erythroxylon* (St Helena redwood) is extinct in the wild and only found in botanic gardens, and *Trochetiopsis melanoxylon* (dwarf ebony) is fully extinct, represented solely by herbarium specimens.

The St Helena ebony's wild population was reduced to a mere two plants, living a marginal existence on a cliff face charmingly known as the Asses Ears. Experiencing *T. ebenus* in the cool shady glasshouse of a botanic garden only accentuates the feeling of loss emanating from their conservation story. Wiry and upright, *T. ebenus* is an elegant architectural shrub adorned with joyful, broad white flowers and a tactile, felty bronze underside known as the indumentum to the leaf and stems.

Thankfully, horticulture can take two plants and make many. Initial cuttings were taken from the survivors and distributed to botanic gardens across the world, slowly building up a substantial ex-situ population. Cuttings are an effective method for increasing plant stocks, but the resulting specimens are clones of their parents, genetically identical and without the variety that may make them resistant to future threats. This is where Kew's expertise with seeds became vital. Finding novel methods to penetrate *Trochetiopsis*'s hard seed coat and to break dormancy allowed a complementary propagation method to bring quality to the quantity of the previous conservation effort. Propagated dwarf St Helena ebony plants are now being replanted on the island, although this initiative is only effective in combination with programmes to remove invasive plants and grazers.

St Helena became ecologically unbalanced through human colonization, but simply "wilding" the island will not restore its pre-human biodiversity and complexity; the balance has tipped too far. It is our responsibility as humans to conserve the island's fragile endemic species using scientific observations to inform our interventions.

J. N. Fitch. del et lith.
Vincent Brooks Day & Son Imp

Finished
Too dark
31. 8.
1809
Finished
Folium seminale

OPPOSITE: *Redwood* (*at B.*) (*Trochetiopsis erythroxylon*) by William Burchell from his *Saint Helena Journal*, 1806–10.

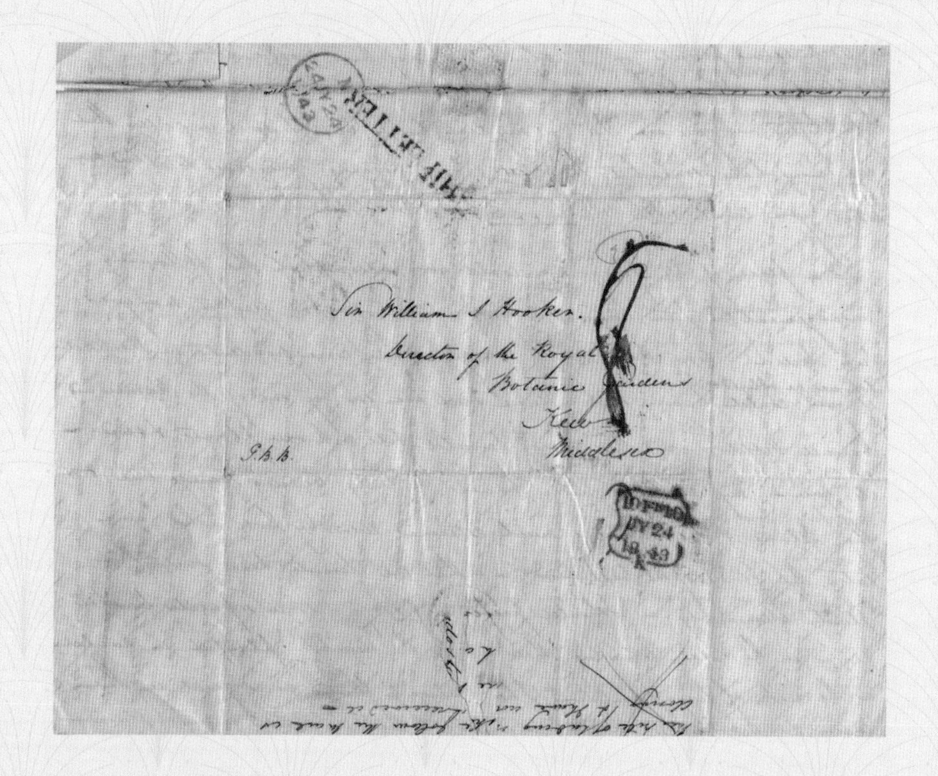

Sir William J Hooker.
Director of the Royal
Botanic Garden
Kew
Middlesex

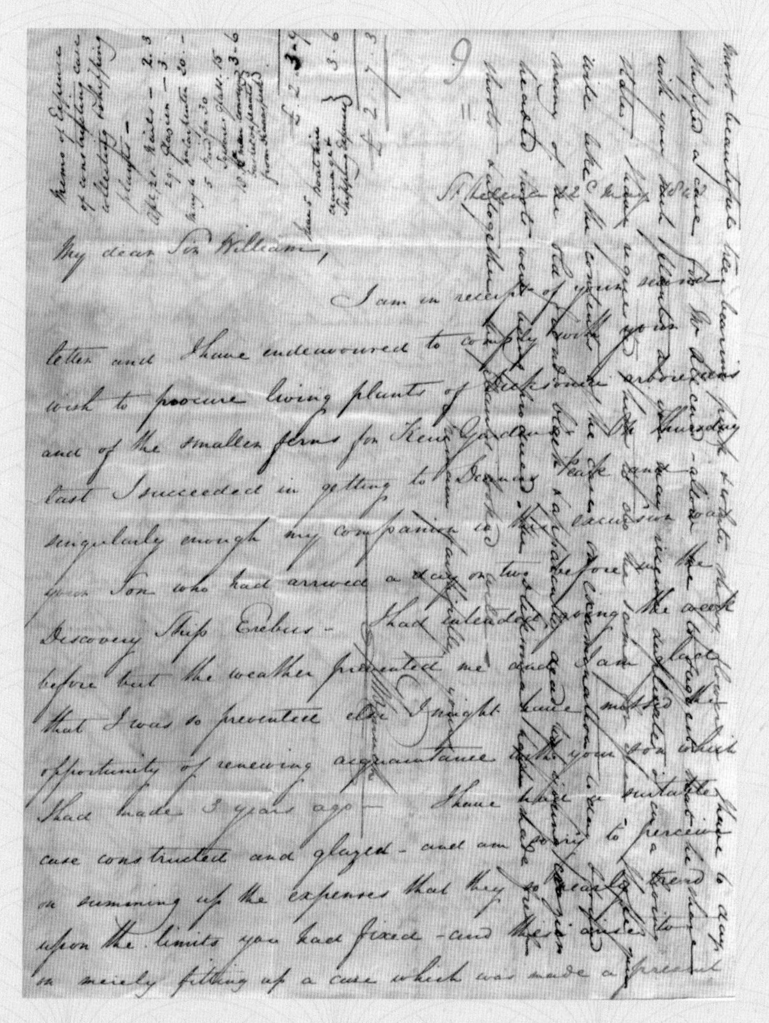

St Helena 22 May 1843

My dear Sir William,

I am in receipt of your second letter and I have endeavoured to comply with your wish to procure living plants of [illegible] and of the smaller ferns for Kew Gardens. On Thursday last I succeeded in getting to Diana's Peak and singularly enough my companion in the excursion was your Son who had arrived a day or two before in the Discovery Ship Erebus - I had intended going the week before but the weather prevented me and I am glad that I was so prevented else I might have missed the opportunity of renewing acquaintance with your Son which I had made 3 years ago - I have had a suitable case constructed and glazed - and am sorry to perceive on summing up the expenses that they so nearly trench upon the limits you had fixed - and these arise from merely fitting up a case which was made a present

Letter from Mr Bennett from St Helena, to Sir William Jackson Hooker, Director of Kew, 22 May 1843.

Bennett writes to Hooker with updates on his plant collecting and exploration of the island, having recently managed to make it to Diana's Peak. He is shipping back various plants to Kew including nine plants of *Melhania melanoxylon* (*Trochetiopsis melanoxylon*).

MISTLETOE

Viscum album

Mistletoe is inexorably intertwined with our folklore and culture. It's a plant of contradictions, symbolizing both love and death – it's a parasite under threat due to rapid loss of its host's environment, and a plant that many associate with their first kiss, but that is, rather unromantically, spread by bird faeces.

A hemiparasite, mistletoe draws nutrients from the vascular system of its host plant, but is capable of photosynthesis. An abundance of mistletoe can fatally weaken elderly trees and occasionally cutting back the parasite is part of traditional orchard management. Consistently forming neat globes up to a metre in diameter, mistletoe is an evergreen plant with notably sticky white berries (or drupes, to use the botanically correct term) borne from autumn to spring.

For the most part, mistletoe is dispersed by birds smearing its berries onto branches or the berries passing unaffected through the birds' digestive tracts. Young plants germinate on the bark of their host branch, with the hypocotyl (embryonic stem) penetrating the surface as it grows. Mistletoe becomes woody with age and can dominate the crown of the host tree.

The plant plays a sinister role in Norse mythology. The goddess Frigg is desperate to ensure no harm befalls her beloved son Baldur, so she asks all living things to promise not to harm him. All living things but one, that is: the humble mistletoe. The mischievous Loki spots this oversight and passes a mistletoe arrow to Baldur's blind brother Holder who, during a game, unwittingly drives the arrow into his brother's heart, killing him.

Mistletoe has permeated other folk traditions, from Anglo-Saxon to Greek to Druidic, with the themes of love, fertility and peace recurring in each culture. The tradition of kissing under the mistletoe is also linked to several countries, the plant's evergreen vitality and associations with fertility underpinning its romantic credentials.

ABOVE: Woman carrying mistletoe through the snow. Wellcome Collection.

OPPOSITE: *Viscum album* from Köhler, *Medizinal-Pflanzen*, 1887.

Tractatus

Ca. ccccxcv.

Virem Secundum serapionẽ libro
aggregatoris cap̄. virem sunt spe
cies scz. absi et indum q̃d appellat̃
arabicum. ⁊ est rubei coloris. ⁊ est idez q̃d
curcuma. Radix eius affert̃ ex terris ĩ bye
me. ⁊ hz grana sicut grana passule. ⁊ meli
us ex eo est rubeũ h̃ns pauca grana. et ali
ud est q̃d adheret manui. ⁊ habet colorez
viole. ⁊ exterius est rubeum. cuius toxicũ
multum. Et toxicum q̃dem est res pauca
minuta mollis adherens manui qñ tangi
tur vagina eius. et ip̄m q̃dem linit̃ super
morpheam confert ei.

Operationes.

A Virtus virem est calida ⁊ sicca in
principio tercij gradus quod si linit̃ ad
morpheam confert B Viriditas in
uenta super aquam que arabice vocatur
batalep. ⁊ in lingua hyspanica mulsas se
cundum Auerroim secundo Colliget. est
calida ⁊ humida in secundo.

Ca. ccccxcvj.

Viscus latine ⁊ grece. Arabice. da
balch vel hele vel hisos. Serapiõ
li. aggre. ca. dabalch aucto. dyas.
Dabalch. i. viscus. Melior ex eo est nouus
qui similis est in colore suo colori porri.
Exterius vero est subrufus sine asperitate ⁊ si
ne fursureitate. ⁊ fit ex fructu rotundo par
uo qui nascit̃ in arbore glandium. ⁊ folia
eius sunt sicut folia arboris bussi. fructus
h̃ iste cõtundit̃ ⁊ abluit̃. deinde coquit̃ in
aq̃. Sunt autem quidam qui masticãdo
fructum ip̄m faciunt viscum. reperit̃ etiã
in arbore piro⁊ ⁊ malo⁊. ⁊ in multis alijs
arboribus. ⁊ aliqñ in radice arborum ali
quarum pua⁊. ⁊ latine vocatur hec plãta
vissago. Ego aũt vidi in mõte Nabazat
hõies qui faciunt ex arbore magna q̃ nas
citur apd̃ eos ⁊ apud ip̄os vocat̃ tarabel
la. ⁊ latine vocat̃ Caprifoliũ. Accipiũt q̃
dem ramos ipsius ⁊ excoriant eos facien
tes fasciculos. ⁊ infundunt in aq̃ et dimit
tunt sic aliquo ex mensibus. deinde ꝯtun
dunt ꝯtusione forti. ⁊ fit inde viscus bonꝰ
Et virtus eius est similis qui subtiliat at
trahit ⁊ dissoluit. q̃r habet v̉tutem resolu
tiuam. Dias. cap̄. viscus s̃m translatõem
nostram. Viscus in ramulis nascẽs a quo
dam animali volatili. scz turdella viscat̃.
Nam illud si morat̃ in ramulis arborum
glutinat̃ ⁊ inde nascunt̃ v̉gule plurime. ⁊
fit viscus q̃ sp̄ viridis inuenit̃. Est aũt opti
mus qui in arbore quercus crescit. et q̃ est
mundus subuiridis. ⁊ si extendat̃ non cor
rumpit̃ sed vt membrana extenuat̃.

Operationes.

A Gal. vj. sim. far. ca. de visco. Sb̃a vi
sci est ꝯposita ex aquosa substãtia ⁊ aerea
⁊ de his ambabꝰ est ĩ eo multũ. ⁊ ex terrea
pauca. ⁊ in sapore eius est acuitas plusq̃
amaritudo. B Et v̉tus eius est q̃ ca
lefacit nõ t̃n cito qñ apponit̃ corpi. sz post
q̃ facit morã in corpe sicut facit tapsia. et
nos quidẽ diximꝰ in his q̃ p̄cesserũt. q̃ hec
res ĩuenit̃ in medicinis qua⁊ v̉tus est que
calefacit cũ q̃⁊ calefactõe est hũiditas su

Mistletoe engraving from *Hortus Sanitatis*, 1511.

Hortus Sanitatis is a natural history encyclopedia from the sixteenth century, which also included mythical creatures.

Replicating the action of a berry passing through a bird's digestive system and locating a perfectly receptive section of branch for germination is, unsurprisingly, a hit-and-miss affair. Even the Royal Botanic Gardens, Kew have struggled to intentionally cultivate mistletoe, with seed-rubbing trials on the branches of the gardens' apple (*Malus*) species that only saw occasional germination occurring. A cool, humid climate is vital for establishment and young seedlings are incredibly vulnerable to drought.

Mistletoe parasitizes a range of deciduous temperate European trees, including lime, willow, poplar, and apple, which is the most common host. Great Britain's ancient orchards have become inextricably linked to mistletoe: the craggy cider trees with gnarled branches are festooned with the emerald spheres of their parasitic guests, underpinned by carpets of wild flowers.

The traditional orchard, a low-intensity form of land management, is in sharp decline, with 90 per cent of the UK's traditional orchards lost since the 1950s. Changing land use and property development have driven this drop, and the result is the depletion of a landscape that is beautiful, biodiverse and productive.

The plant's dependence on declining ancient orchards means mistletoe, too, is disappearing from its traditional British heartlands: the counties of Herefordshire, Somerset, Worcestershire and Gloucestershire. Supporting conservation organizations such as The Orchard Project or buying traditional apple varieties directly from growers helps conserve a wonderful habitat and one of our most treasured, enigmatic plants.

RIGHT: *Australian Sandal Wood with Mistletoe and Emu Wren, West Australia,* by Marianne North, 1880.

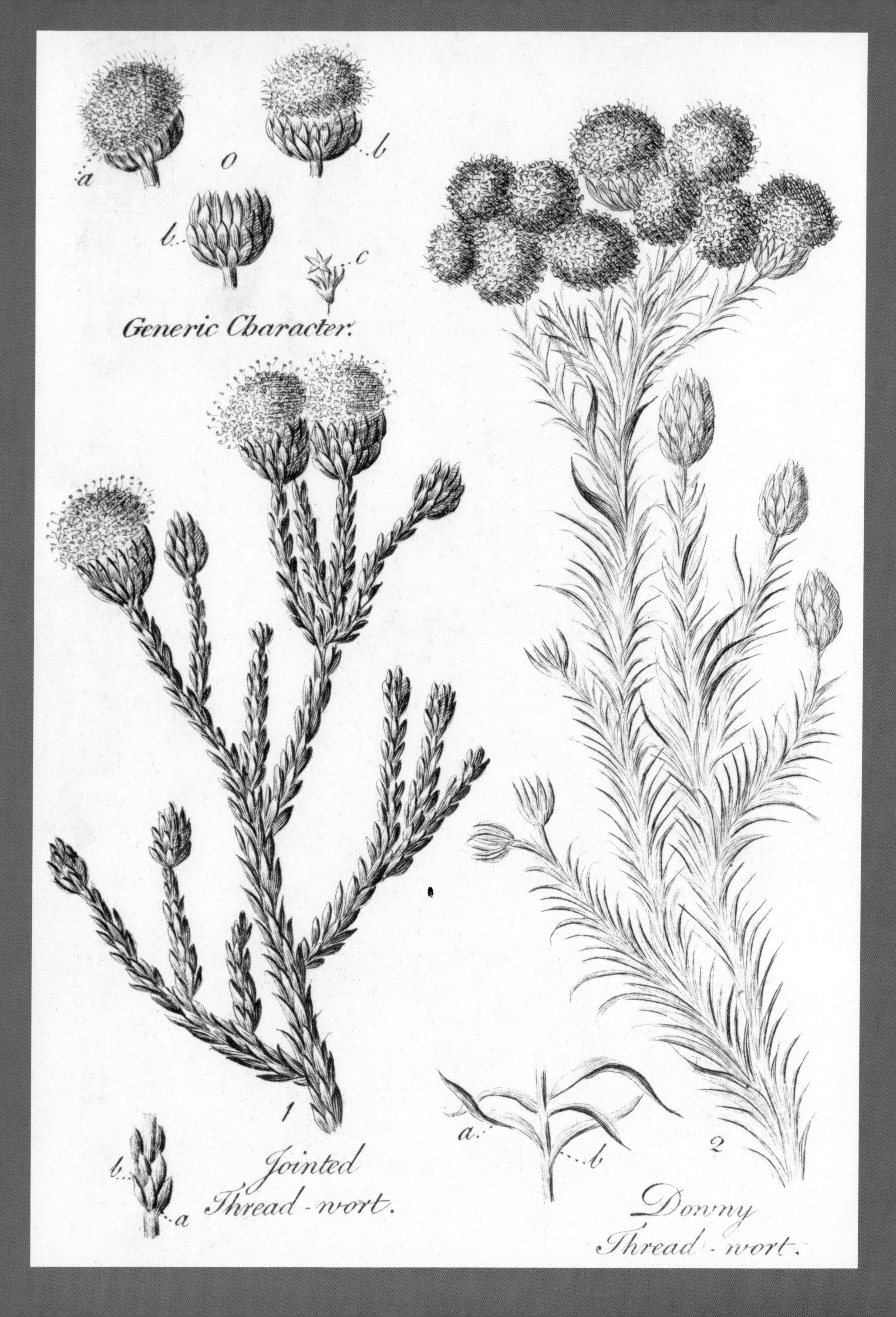

a
o
b
b.
c
Generic Character.
1
b.
a
Jointed
Thread-wort.
a.
b
2
Downy
Thread-wort.

40

MULANJE CEDAR

Widdringtonia whytei

A genus of four coniferous species only found in southern Africa, *Widdringtonia* is a prized plant for connoisseurs of rarity, its exclusivity heightened by precise cultivational requirements. The species in this genus are renowned for their fragrance, fine-grained timber and termite resistance, and these values have been its downfall in the wild. The decline of the Mulanje cedar (*Widdringtonia whytei*) is an unsettling example of a species heading to critical endangerment, despite concerted conservation efforts.

Widdringtonia plants, especially *Widdringtonia wallichii* (Clanwilliam cedar), add a stately presence to tree collections, with mature specimens reminiscent of Lebanese cedars, albeit with softer, cypress-like foliage. Establishing a specimen in UK conditions is considered a horticultural triumph: all species are borderline hardy and are killed by hard frost. Sheltered gardens and arboretums tempered by the moderating effects of the Gulf Stream, especially in Cornwall and Ireland, are the best places to find cultivated *Widdringtonia*.

The plight of the Mulanje cedar means botanic gardens, arboretums and nurseries may be the only places to experience this beautiful, valuable tree. Their range is extraordinarily limited, their sole wild location being Mulanje Massif mountain in Malawi, southern Africa's tallest peak. Living at an altitude between 1,800 and 2,500 metres and favouring shallow acidic soil, this tree is inherently vulnerable to changing land use because of its narrow range. Although adapted to the natural bush fires that maintain succession from forest to grassland and back,

OPPOSITE: *Widdringtonia nodiflora* (as *Brunia nodiflora*) from John Hill, *The Vegetable System*, 1714–75.

RIGHT: Herbarium specimen of *Widdringtonia whytei* collected by J. D. Chapman and I. H. Patel in Malawi in 1981, held at the Royal Botanic Gardens, Kew.

2
5
4
3
1

the Mulanje cedar is slow to regenerate. Some fire-tolerant trees re-sprout from burnt trunks, but this tree relies on seedlings emerging from the charred scrub.

The forces pitted against the Mulanje cedar epitomize almost every threat facing the world's rare plants: invasive new neighbours, exotic pests, uncontrolled wildfires, mining and felling. As the national tree of Malawi, this cedar has a prominent place in the national conscience; its first conservation reserve was declared in 1927. Felling should be a strictly licensed activity – and the only way to access timber that's highly desirable for building boats, houses, furniture and walking sticks – but economic forces have overwhelmed legislation and conservation.

Clearance of the Mulanje cedar populations started over a hundred years ago and has dramatically accelerated in the last decade. A distribution that once occupied over 600 square kilometres is now reduced to a mere 16 square kilometres, earning an IUCN Red List rating of "Critically Endangered". Equally desperate is the viability of the remaining trees. There are no mature adult specimens left and while there's limited evidence of regeneration, the remaining population is barely functional. Attacked by exotic aphids, outcompeted by invasive bracken and pressurized by wildfires, this weakened population has little resilience to further threats, a national symbol fading fast.

Can the Mulanje cedar be saved? A determined initiative is trying to change the ending to this predictable script. Botanic Gardens Conservation International (BGCI) offers global coordination and expertise to fulfil its mission of stopping species becoming extinct. Working with Malawian conservation and forestry bodies, Mulanje cedar nurseries are raising young plants by the thousand. After achieving the immediate aim of ensuring that Malawi's national tree doesn't become extinct in-situ, the wider ambition is to adapt the economy this tree has driven. Commercial Mulanje cedar plots could shift use from wild exploitation to controlled cultivation, supported by exploring wider commercial uses, including the potency of the tree's essential oils.

Reintroduction to the Mulanje Massif has a romantic appeal, but is only viable with both a dramatic shift in local economy and concerted suppression of the species' myriad threats. A tree worth breaking the law to fell cannot be protected by legislation alone.

OPPOSITE:
Original illustration of *Widdringtonia whytei* by Olive H. Coates Palgrave, produced for *Trees of Central Africa*, 1957.

ABOVE:
Herbarium specimen of *Widdringtonia whytei* collected by L. J. Brass in Malawi in 1946, held at the Royal Botanic Gardens, Kew.

a
g
f
c
b
l
k
i
h
e
d
Brunia nodiflora

OPPOSITE: *Widdringtonia nodiflora* (as *Brunia nodiflora*) from Johann Christoph Wendland, *Collectio plantarum, Sammlung -ausländischer und einheimischer Pflanzen*, 1805–19.

ICONES CENTURIÆ PRIMÆ. 10

CUPRESSOPINULUS
CAPITIS BONÆ SPEI.

*

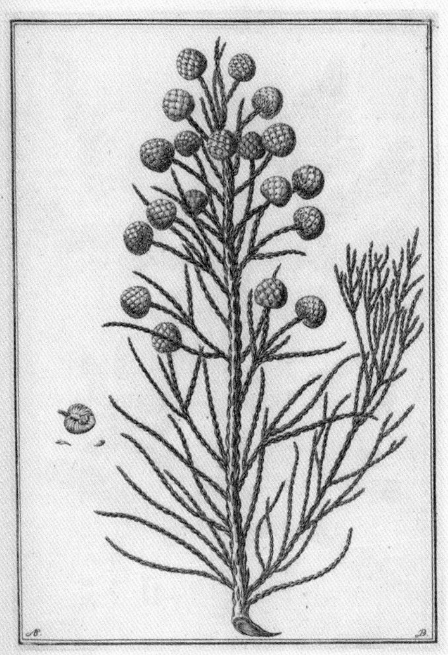

C 2

RIGHT: Title page and pages detailing *Cupressopinulus* (*Widdringtonia nodiflora*) from Jakob Breyne, *Jacobi Breynii Gedanensis Exoticarum aliarumque minus cognitarium plantarum centuria prima*, 1678.

22 JACOBI BREYNII PLANTAR.

ſimiles, ſtaminulis pallidis conſtantes, erumpunt. His deciduis Globus ille, in *Conum* fuſcum, uncialem & majorem ſquamis in extremitate ſuâ albidis & lanuginoſis, evadit.

Reperitur quoq; Fruticis hujus coniferi ſpecies major, Foliis majoribus, cujus Fructus in icone ſeorſim ſculptus. De quâ mihi Anno M. DC. LXIV. ramus ſiccus â *Domino Doctore Huyberto*, & Anno M. DC. LXX. alius â *Domino Doctore Matthæo Slado* communicatus.

ILLUSTRISSIMUS etiam DOMINUS à BEVERNINGK antè tredecim annos ramo ex Africâ, mediam inter hunc & illum deſcriptum faciem obtinente potitus: qui mihi, niſi varietas ſubſit ratione loci, Naturæ luſus videtur.

CUPRESSO-PINULUS CAPITIS BONÆ SPEI.

Cap. X.

C*Upreſſo-pinulum* Arborem quandam voco Africanam, ſemper virentem, & ut mihi relatum pumilam. De quâ *Rami* aliquot perelegantes ILLUSTRISSIMO DOMINO à BEVERNINGK Anno M. DC. LXIII. ex Africâ miſſi, ob raritatem & peregrinam formã, Botanicis omnibus, quibus viſi, curioſis, gratiſſimi. Qui *Virgæ* ſunt, pedales circiter, rotundæ, pennæ craſſitiem ſcriptoriæ penè attingentes; ligno ſubalbido; cortice rufeſcente ex cinereo

EXOTICAR. CENTURIA I. 23

nereo tectæ, *foliolisque* anguſtiſſimis, triangularibus ac mucronatis, Cupreſſi inſtar ramulorum, ſquamatæ. His verò Virgis, ramuli undique, duas tresve uncias longi, graciles valdeque flexibiles adnati; qui *foliolis* Ericæ vulgaris perſimilibus, decuſſatim poſitis, densè veſtiuntur. Quilibet ramulorum ſuperiorum, *Strobilo* (modò uno, modò pluribus) exiguo, orbiculari, lanuginoſo & albo, juxta ramulum planiuſculo foliolisq; firmiter adhærentibus ſubtiliſſimè ſcandulato, onuſtus; formâ & magnitudine Fragorum, ſquamulis autem innumeris, intùs villoſis ac argenteo nitore ſplendentibus, in anteriore parte foliolum viride monſtrantibus, arctiſſimè compacto: cujus ſquamæ, ut autumo, proceſſu temporis dilatantur. Lignum & cortex rami mei ſaporem ſalſum (ex aquæ marinæ fortaſſis aſperſione) & nonnihil adſtringentem; ſed foliola cum ramulis ingratum, ſubaſtringentem, cum tantillo amarore præbent.

CHAMÆLARIX, SIVE CHENOPODA MONOMOTAPENSIS.

Cap. XI.

ALunt etiam juxta Bonæ Spei Promontorium deſerta *Fruticem echinatum* quendam, frequentiſſimis ramis undique piloſis divaricatum: qui per interſtitia brevia, invicem ambiuntur, à viridibus, Laricis in modum in acervos

GLOSSARY

Abiotic – not living. Abiotic environmental elements include climate and soil conditions.

Alkaloid – a plant-derived chemical, often bitter or toxic with a range of medicinal applications.

Allelopathy – suppression of competition by certain plants through exudation of chemicals and physical exclusion.

Anthropocene – an epoch where human influence, more than any other, has influenced climate and global speciation.

Aphid – a sap-sucking insect that can be an agricultural and horticultural pest.

Archipelago – a group of islands within a defined geographic area, generally under one country's control.

Atropine – a plant-derived chemical with extensive medical applications.

Biodiversity – the variety and richness of species on earth.

Bioprospecting – the search for valuable products in plants for use in medicine, fibres, fuels and other commercial applications.

Carnivorous plants – a plant either partly or fully dependent on animal matter for nutrition.

Coniferous – cone-bearing trees with "naked" seeds and needle-like leaves. Most coniferous trees are evergreen.

Conservation – the science of reducing threat to endangered species.

Cultivars – a novel plant form raised in cultivation.

Desiccation – the process of drying seeds to prepare them for dormancy in seed banks, prior to freezing.

Dieback – the death of growing parts of a plant, a symptom of ash dieback.

Dioecious – a plant species with separate male and female plants.

Distribution – the naturally occurring wild range of a plant.

DNA sequencing – a method for determining a plant's unique genetic code and understanding its relationship to other species.

Dormancy – a stable state under which a seed can be stored for extended periods of time. In dormancy, seeds do not use their starch.

Ecological Horticulture – the adoption of ecological principles to establish and maintain a designed plant community.

Elaiosome – an oil-rich seed structure.

Endemic/Endemism – a species that only occurs in one specific location.

Epicormic growth – shoots that emerge from previously dormant buds in a tree's bark.

Ethnobotany – the study of connections between plants and people.

Ex-situ conservation – conservation of a species away from its wild distribution.

Fronds – the photosynthetic organ of a fern, equivalent to a leaved stem in a higher plant.

Genus – a taxonomic unit between the ranks of species and family.

Germination – the point at which a seedling begins to sprout from a seed.

Gigantism – a tendency for plants and animals to grow to an unusually large size as a consequence of specific biotic and abiotic conditions.

Habitat – a cumulative set of biotic and abiotic conditions within a defined space that can support life.

Hapaxanth – a plant that flowers once in its lifetime, sets seed then dies.

Hemi-parasites –"half parasite" – a species that relies on its host for a vital resource or function, but does not fully exploit or weaken it.

Herbicide – a chemical formulated to kill plants.

Horticulturist – a professional gardener.

Hybrid – a cross between two taxonomically distinct plants.

Hybridization – the process of creating a hybrid. This can happen spontaneously in the wild or under human cultivation.

Inflorescence – a compound flower.

In-situ conservation – conservation of species within its wild distribution.

Indigenous species – plant or animals occurring within their wild geographic distribution.

Invasive species – a species outside its wild distribution, growing without biotic or abiotic constraints to outcompete indigenous species.

IUCN – the International Union for the Conservation of Nature.

Lignin – a hard woody tissue found in trees and shrubs.

Monocotyledon – a group of plants with single seed leaves and parallel leaf veins, amongst other characteristics. This group includes grasses, palms and orchids.

Monoculture – extensive and repetitive growing of a single species crop.

Morphology – the physical characteristics of a plant, its shape.

Mucilage – a mucus-like substance produced by plants.

Nomenclature – naming.

Orthodox seeds – seeds that can be banked using a conventional approach of desiccation and freezing.

Parasites – a range of life forms that derive all or part of their nutrition through a host species, often to the host's detriment.

Photosynthesis – the biochemical process plants uses to make sugars from sunlight and carbon dioxide, which are then converted to energy.

Phylogenetic tree – a diagram showing the relationship between different species within an evolutionary context. Phylogenetics identifies common ancestors and point where species branch away, becoming more distinct.

Phytochemical – a chemical synthesizes by plants.

Pollinator – an insect or animal that distributes plant pollen from one plant to another, often inadvertently when in search of nectar.

Population – an ecologically functional group of a single species existing within a defined geographical area.

Proboscis – the sucking mouthpart of an insect, often adapted to feed on a specific plant's nectar.

Progeny – offspring

Propagation – the practice of duplicating or replicating plants by seed or vegetative means (cuttings, layering, grafting).

Raceme – a type of inflorescence with equally spaced flowers borne on equal length flowering stems.

Re-wilding – a movement that seeks to reverse intense human management of land for the benefit of nature.

Recalcitrant – seeds that will not become dormant under desiccation and freezing and expire. The opposite of orthodox and a major seed banking challenge.

Regeneration – the capacity of a plant to repair damage, often associated with post-fire recovery.

Remote sensing – a method using satellites and drones to map plants on a large scale, monitoring a range of factors such as plant health and vegetation coverage.

Resin – a viscous liquid produced by plants, often in response to injury.

Rhizoids – the equivalent of a root structure in lower plants like ferns and mosses.

Seed distribution – the method by which plants move seeds away from them, ensuring their progeny do not compete with them for resources.

Seedling – the germinated emergent form of a plant.

Species – a taxonomic rank below genus and above sub-species.

Spores – the sexually reproductive structure of ferns and fungi.

Stem succulence – a trait found in plants adapted to arid environments to store water within plant tissue.

Stewardship – human land use that is not exploitative or detrimental to biodiversity.

Taxonomy – the science of naming and defining relationships between plants.

GENERAL READING

Bynum, Helen & Bynum, William, *Remarkable Plants That Shape Our World*, Thames & Hudson, 2014.

Cocker, Mark, *Our Place*, Jonathan Cape, London, 2018.

Curtis's Botanical Magazine, continuously published since 1787.

Davis, Aaron P. *et al*, *Coffee Atlas of Ethiopia*, Royal Botanic Gardens, Kew, 2018.

Desmond, Ray, *The History of the Royal Botanic Gardens, Kew* (2nd edition), Royal Botanic Gardens, Kew, 2007.

Dransfield, John *et al*, *Genera Palmarum: the evolution and classification of palms*, Royal Botanic Gardens, Kew, 2008.

Farjon, Aljos, *Ancient Oaks in the English Landscape*, Royal Botanic Gardens, Kew, 2017.

Fortey, Richard, *The Wood for the Trees*, William Collins, London, 2016.

Fry, Carolyn, Seddon, Sue & Vines, Gail, *The Last Great Plant Hunt*, Royal Botanic Gardens, Kew, 2011.

Fry, Carolyn, *The Plant Hunters*, André Deutsch, London, 2017.

Gardiner, Lauren & Cribb, Phillip, *The Orchid*, André Deutsch, London, 2018.

Goulson, Dave, *A Buzz In The Meadow*, Jonathan Cape, 2014.

Harrison, Christina & Kirkham, Tony, *Remarkable Trees*, Thames & Hudson, 2019.

Hay, Alistair, Gottschalk, Monika & Holguín, Adolfo, *Huanduj: Brugmansia*, Royal Botanic Gardens, Kew & Florilegium, New South Wales, 2012.

Mills, Christopher, *The Botanical Treasury*, André Deutsch, London, 2016.

North, Marianne, *Official Guide to the North Gallery* (facsimile of 6th edition), Royal Botanic Gardens, Kew, 2009.

Payne, Michelle, *Marianne North: a very intrepid painter*, Royal Botanic Gardens, Kew, 2011.

Rix, Martyn, *The Golden Age of Botanical Art*, André Deutsch, London, 2018.

Royal Botanic Gardens, Kew, *Marianne North: the Kew Collection*, Royal Botanic Gardens, Kew, 2018.

Ulian, Tiziana *et al*, *Wild Plants for a Sustainable Future*, Royal Botanic Gardens, Kew, 2019.

Willis, Kathy & Fry, Carolyn, *Plants: from roots to riches*, John Murray, London, 2014.

INDEX

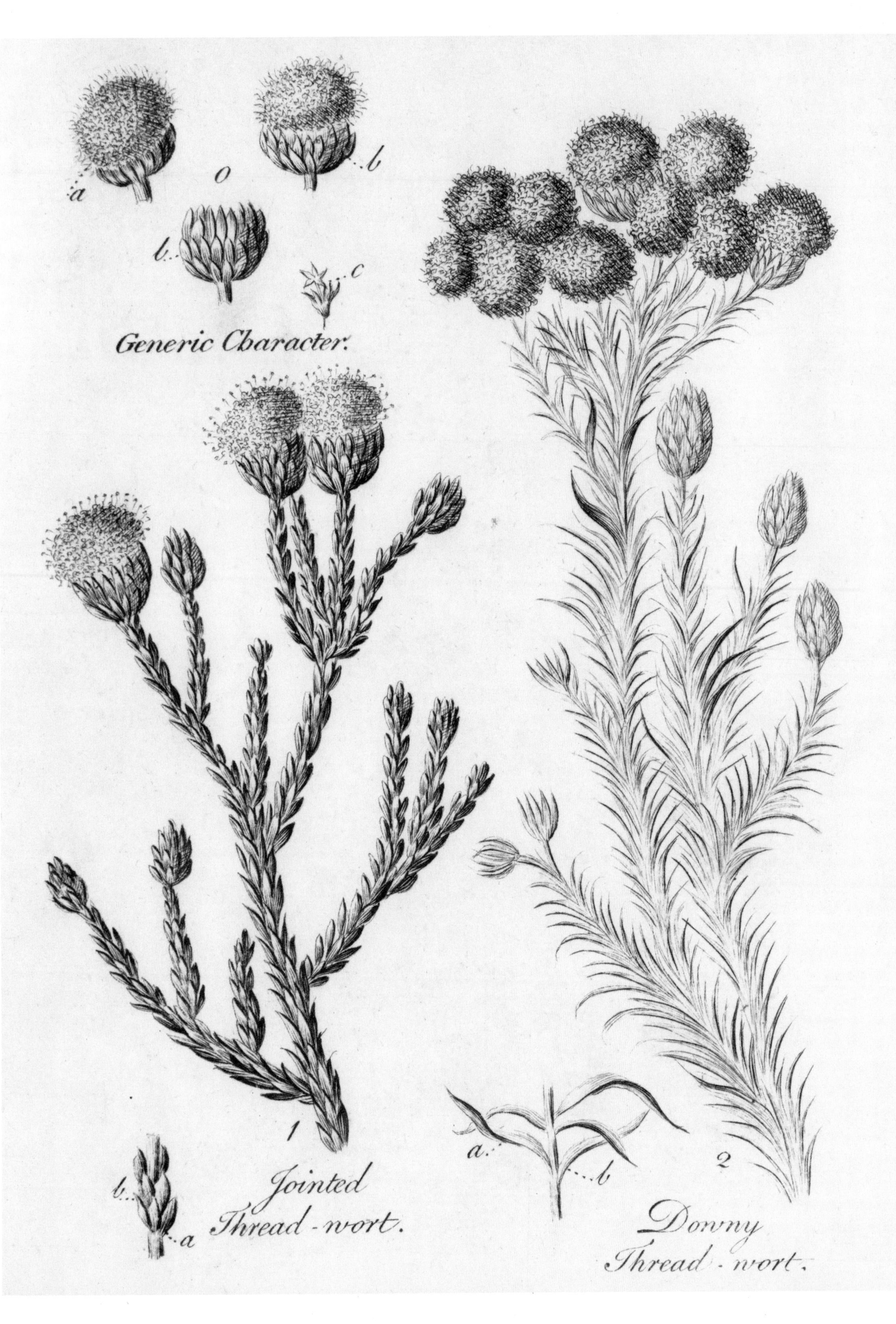

a
o
b
b
c
Generic Character.
1
b
a
Jointed
Thread-wort.
a
b
2
Downy
Thread-wort.

Loranthaceae.

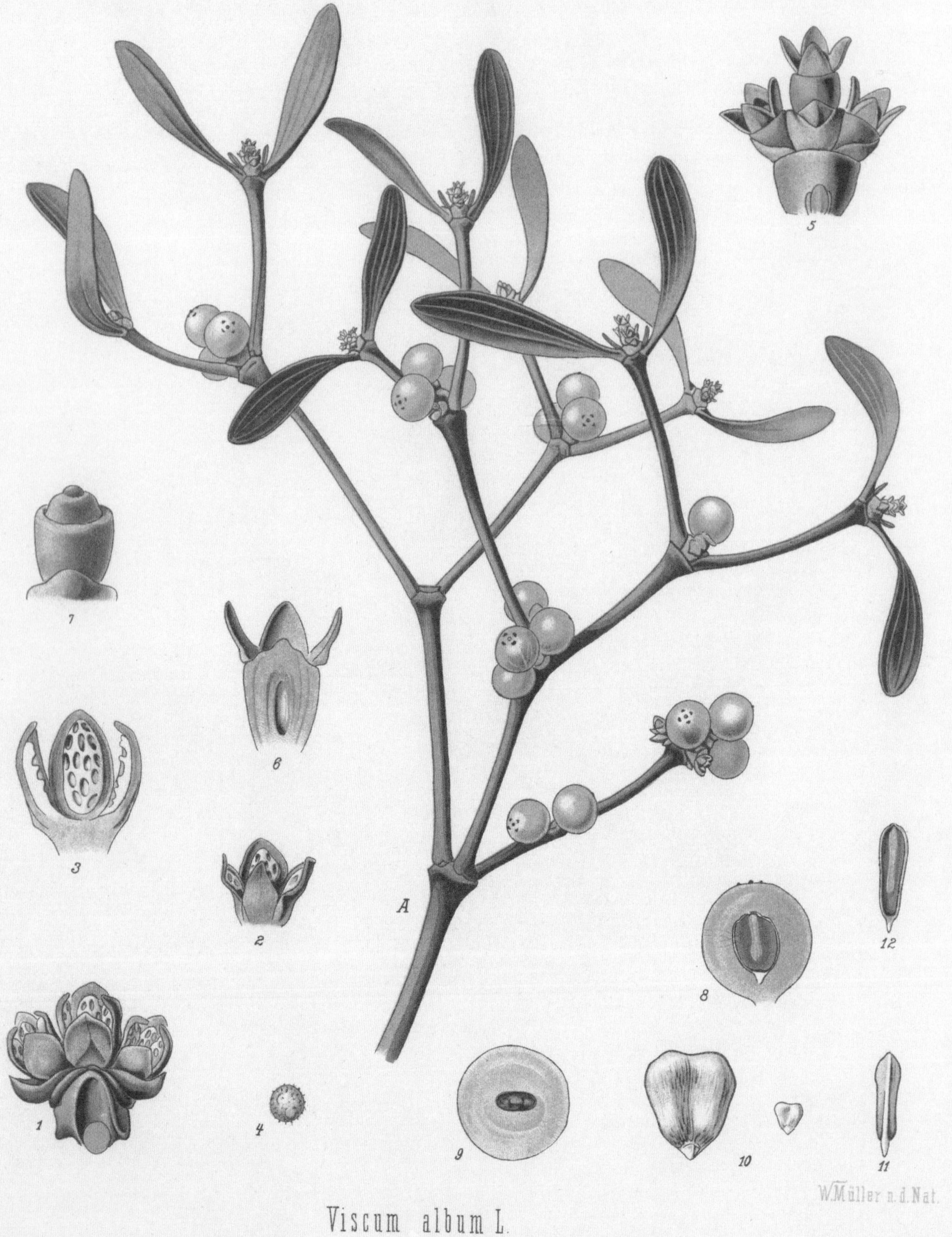

W. Müller n. d. Nat.

Viscum album L.

Syd. Edwards del. Pub. by T. Curtis, St. Geo. Crescent Mar. 1. 1807. F. Sansom sc.

J.R. Guillot. del. *Tecophilaea Cyanocrocus.* Chromolith. J.L. Goffart Bruxelles

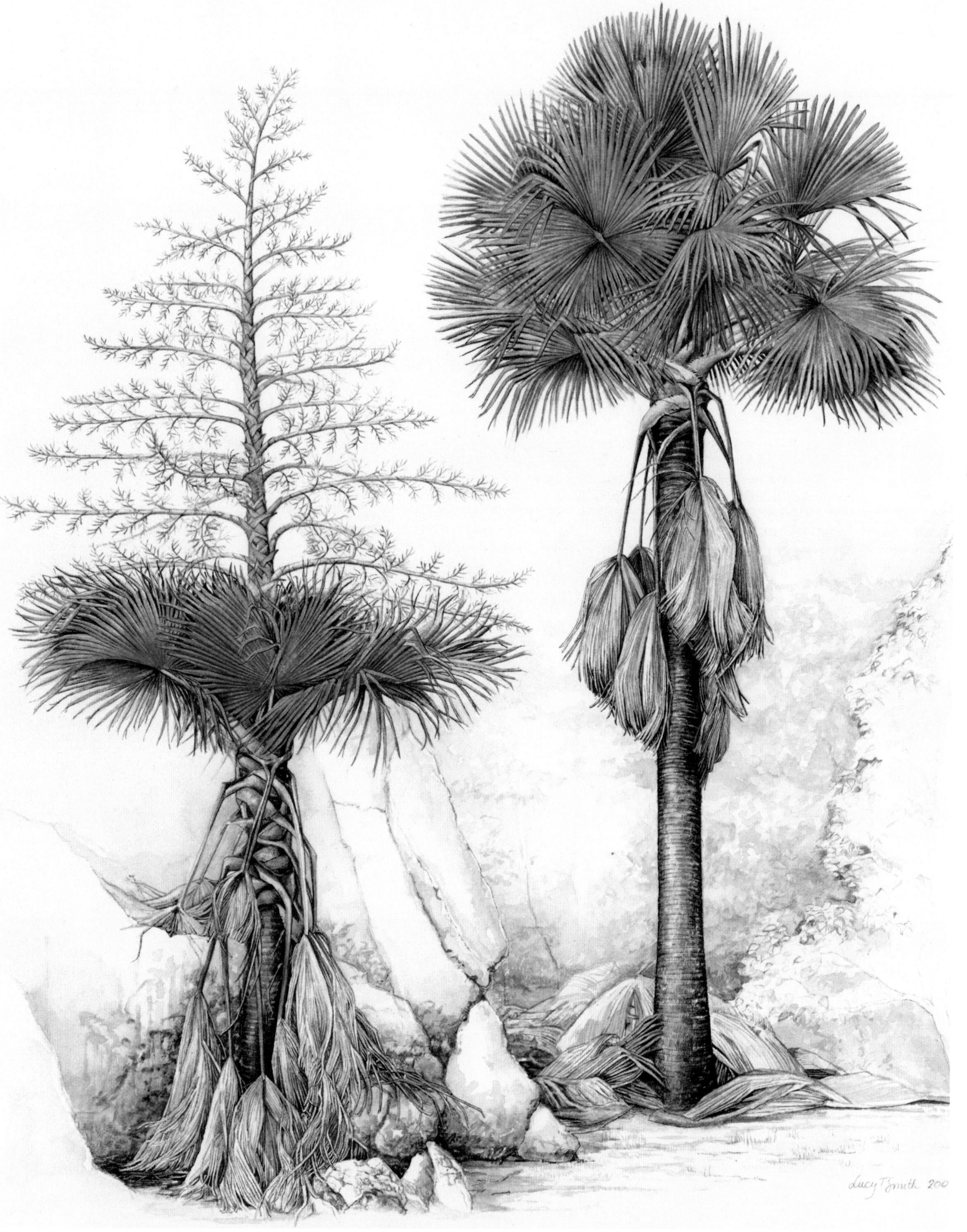
Lucy T Smith 200

L Reeve & C° London

XXI, 5.
34. Cupuliferae.
A
B
1
2
3a
3b
4
5
6
7
W. MÜLLER, GERA.
161. Quercus pedunculata Ehrhart. Stieleiche.

Plate XLVIII.

Pulsatilla

E.H. delt., A.M. Traice, lith. (1. Anemone Pratensis L. / 2. Anemone Pulsatilla L.) Thos. Harrild, imp.

Platanus orientalis.

CYPRIPEDIUM BELLATULUM RCHB. F.

3914.
W. Fitch del.t
Pub. by S. Curtis Glazenwood Essex Dec.r 1. 1841.
Swan Sc.

Plate 690 *Nymphaea thermarum* LUCY SMITH

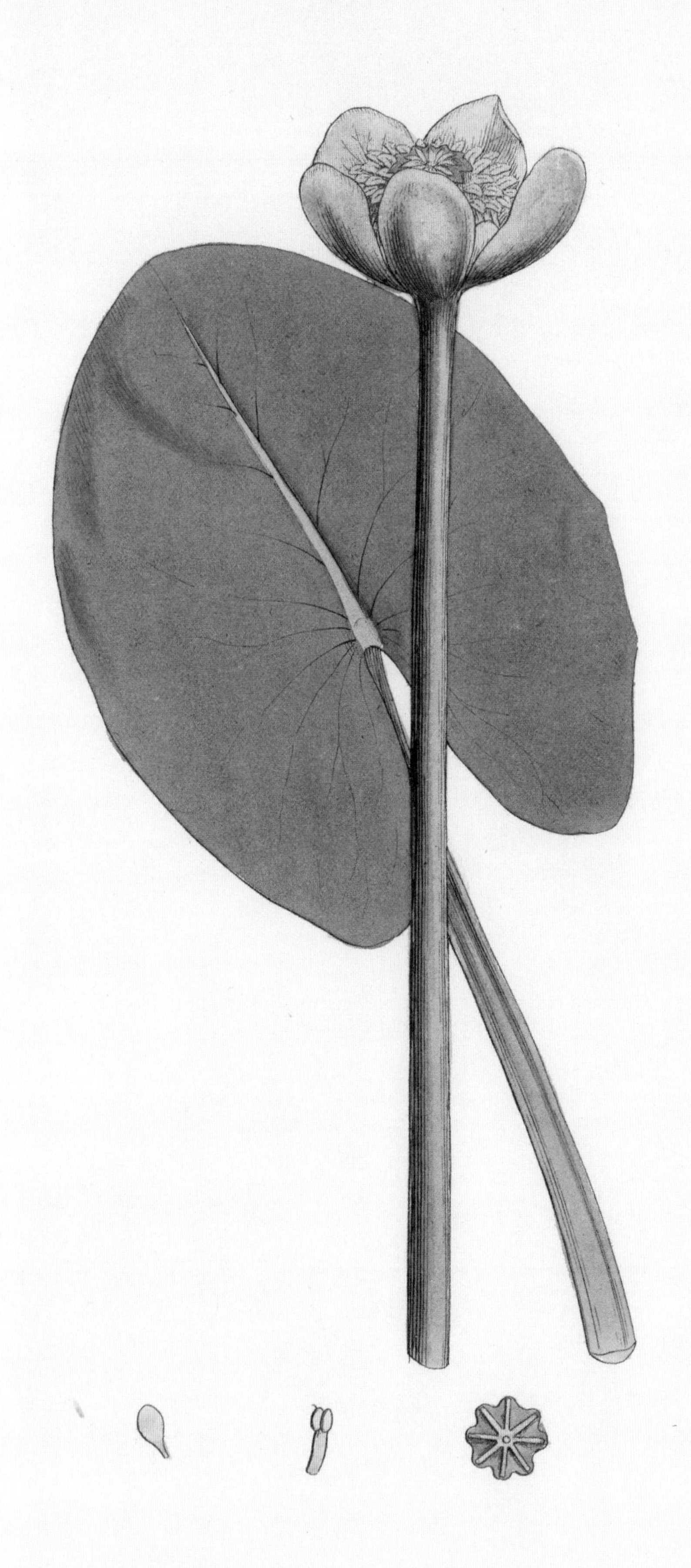

Nuphar pumila. Least Water Lily.

NEPENTHES
ATTENBOROUGHII

Melampyrum cristatum Linné.

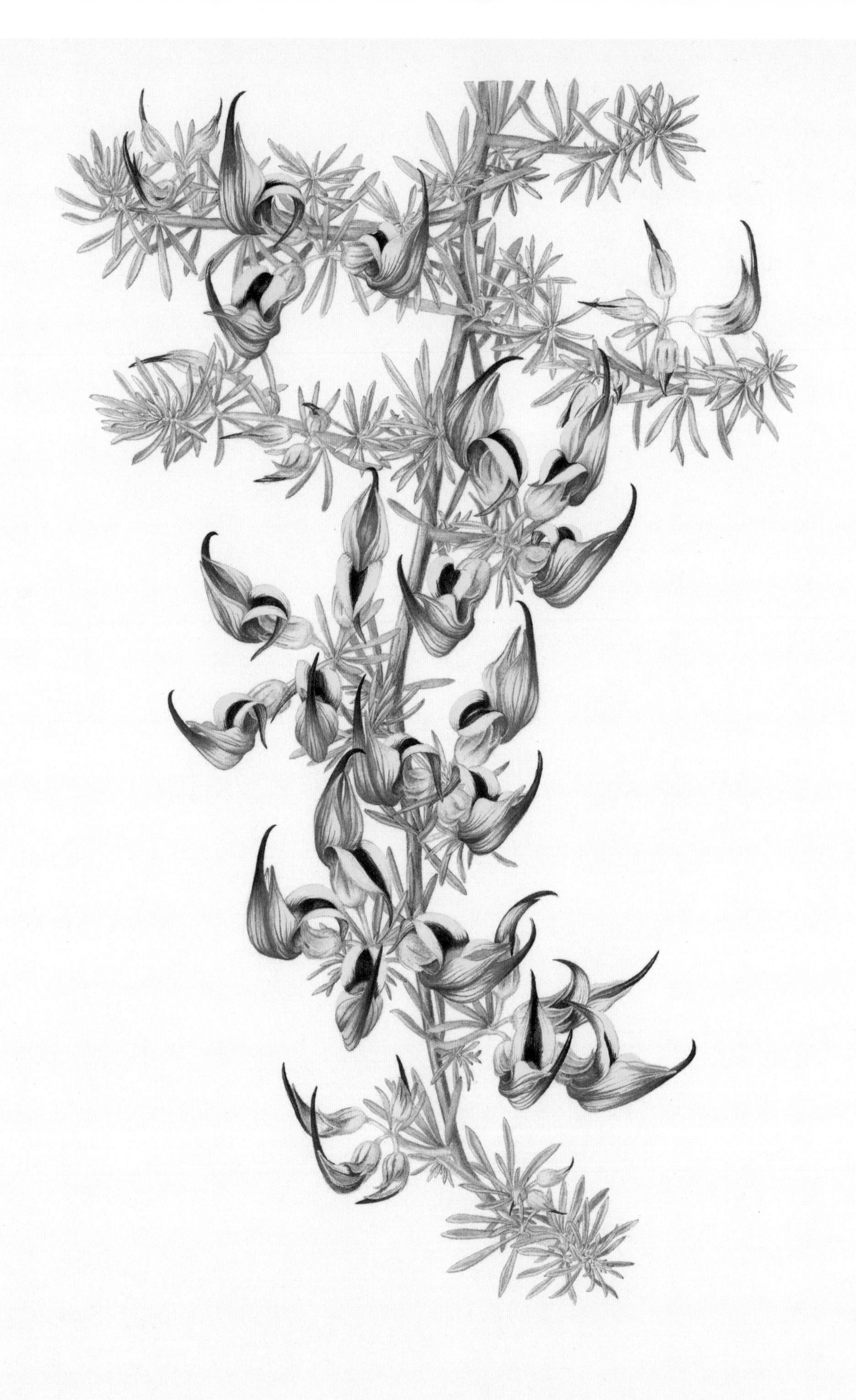

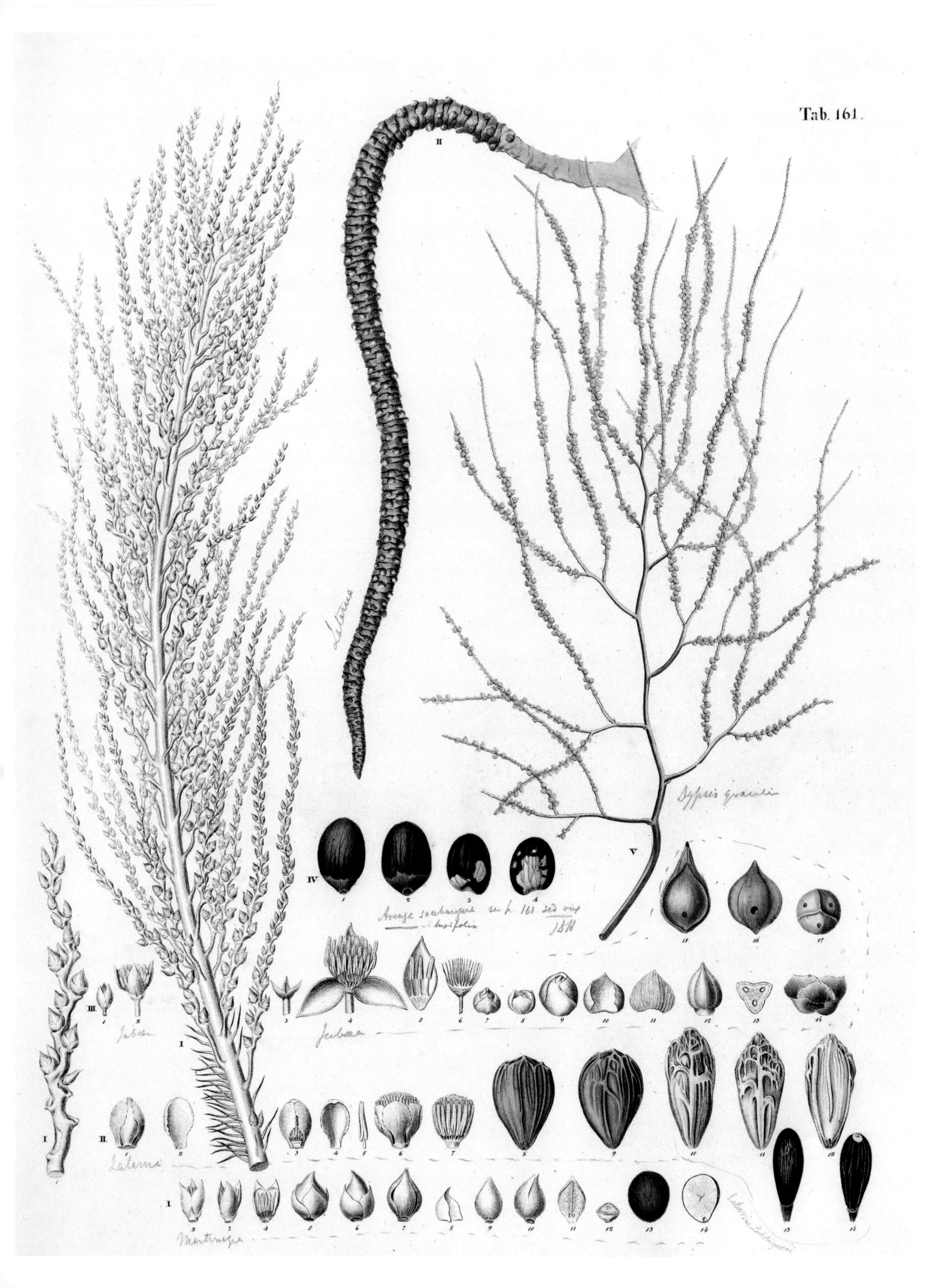
Tab. 161.
II
IV
V
III
I
II
I
I

N. 611.
a
b
c
d
a. Iris Anglica bulbosa flore purpureo.
b. Iris Susianna.
c. Iris bulbosa variegata.
d. Iris vulgaris humilis flore violaceo.
s.

NATURE PRINTING.

Trichomanes radicans.

L. Reeve & C° London.

11

Miss Drake del. Pub by J. Ridgway 169 Piccadilly Feb.y 1 1844 G. Barclay sc

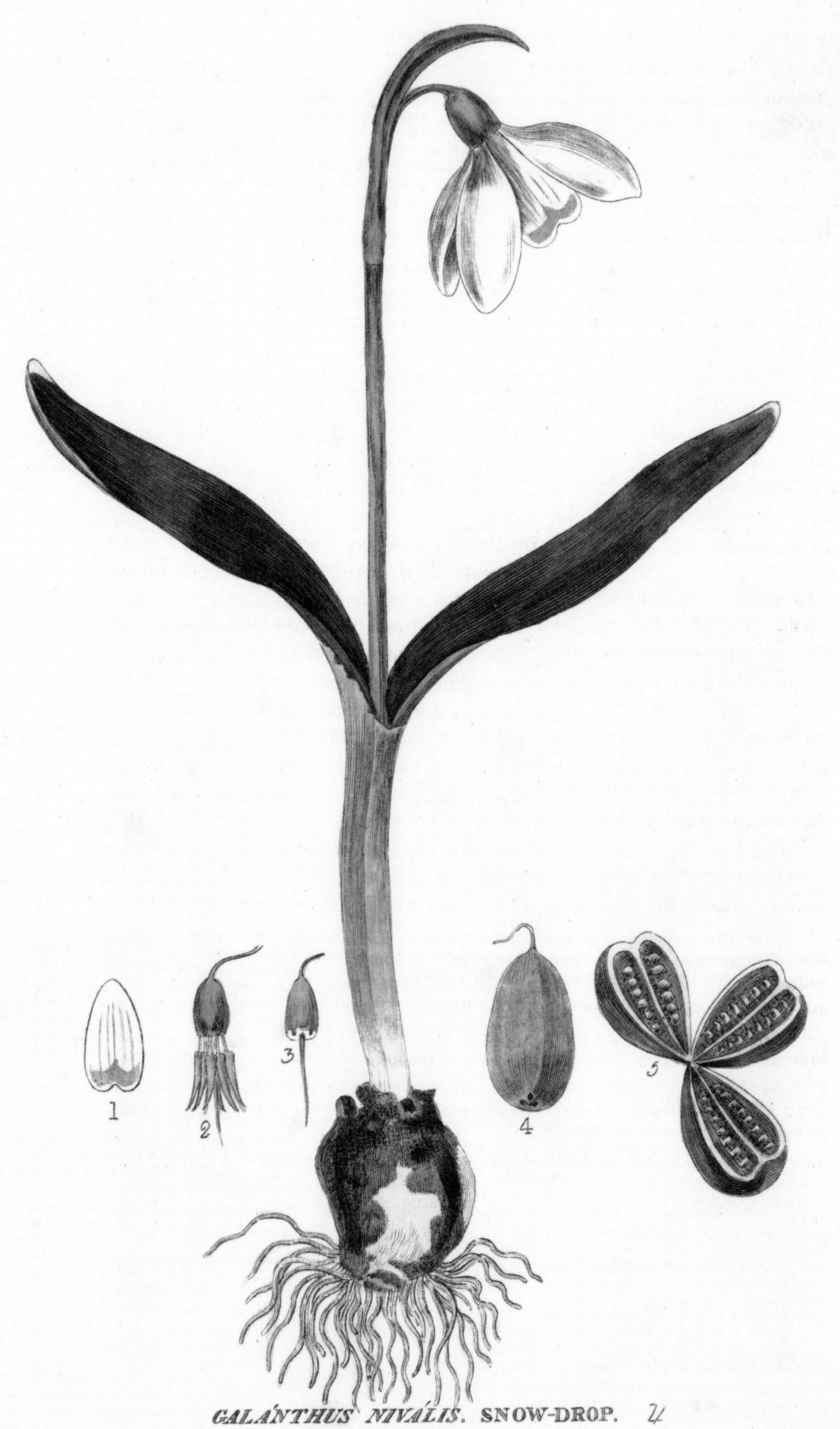

GALÁNTHUS NIVÁLIS. SNOW-DROP. 4

I. Russell. Del. Pub. by W. Baxter. Botanic Garden Oxford. 1833. C. Mathews. Sc.

1. 2. Acacia Ausfeldi Rgl.
3. Fritillaria Meleagris B.

II, 1.
131.
Oleaceae.
B
A
C
2
5
7
6
8
3
4
9
1
534.
Fraxinus excelsior L.
Gemeine Esche.

Eucalyptus Globulus Labillardière.

W. Fitch, del. et lith.
Vincent Brooks, Imp.

Echinacea intermedia Lindl.

7475

Arentine H. Arendsen ad nat. del.

Chromolith. G. Severeyns. Brussel.

IMANTOPHYLLUM MINIATUM.

A. C. van Eeden & Co. Wagenweg N° 91, Haarlem
Editors and Publishers

1775.
Miss Drake. del.
Pub by J. Ridgway 169 Piccadilly July. 1. 1835.
J. Watts. sc.

N. 2218.

Pub. by S. Curtis, Walworth, March 1. 1821.

Weddell Sc.

Senecio Huntii.

Leighton Brothers. Lith.

PLATE 13.

ARAUCARIA IMBRICATA *Pav.*

♄ *Chili.* *Plein air.*

452

Aloe vera Costa spinosa.